Geographical Information System and Crime Mapping

Geographical Information System and Crime Mapping

Monika Kannan and Mehtab Singh

CRC Press is an imprint of the
Taylor & Francis Group, an **informa** business

First edition published 2021
by CRC Press
6000 Broken Sound Parkway NW, Suite 300, Boca Raton, FL 33487-2742

and by CRC Press
2 Park Square, Milton Park, Abingdon, Oxon, OX14 4RN

First edition published by CRC Press 2021

CRC Press is an imprint of Taylor & Francis Group, LLC

ISBN: 9780367359034 (hbk)
ISBN: 9780429342554 (ebk)

Typeset in Times
by MPS Limited, Dehradun

Dedication

To all the individuals we have had the opportunity to lead, be led by, or watch their leadership from a far, we want to say thank you for being the inspiration and foundation for writing this book.

Contents

Foreword

FOREWORD I

In a country where rapid economic development, globalization, population increase, and other societal changes are taking place, conventional ways of crime management based primarily on intelligence and criminal records are becoming increasingly inadequate. It is necessary to aggregate voluminous crime and other related data, to identify patterns and clusters of their distribution, to explore the relationship between crime and related factors, and to assess the effect of crime reduction strategies. All of these can be done efficiently by using GIS mapping, which is a tool for visualization, analysis, and maintenance of the crime database. The authors of this book see great potential for crime reduction with this process aided by artificial intelligence and the Internet.

This book covers such diverse themes as theoretical principles of crime occurrence, GIS, analytical techniques, data processing solutions, information sharing, problem-solving approaches, and map design. It also includes organizational structures for using crime mapping for crime analysts, planners, educationalists, law enforcement agencies, and other stakeholders. It provides a complete start-to-finish coverage of crime mapping, including theory, scientific methodologies, analysis techniques, and design principles.

This book is the product of several years of tireless effort in applying GIS technology to crime mapping with practical concerns and real-world applications. It is hoped that it is used not only by those who wish to reduce crime for the benefit of society, but also by those who intend to explore a new model of a well-managed, crime-free society relevant for a huge democratic country such as India.

Yukio Himiyama
Emeritus Professor, Hokkaido University of Education
President, International Geographical Union

FOREWORD II

Urbanization is fast expanding in India and presently about 31% of the total population resides in urban areas and it is predicted to increase to 50% by 2030. Added to this, slums contain 20% of the population in urban agglomerations. It is also observed that one-third of the population migrates to different parts of the country for seasonal and long-term employment. Different regions of the country also experience calamities such as floods, cyclones, droughts, etc., that have major impacts on the economy of the affected population. Due to the above reasons, studies show widening economic gaps between the rich and the poor resulting in an increase of crime.

To better understand this complex problem of crime in a country which has a wide diversity of ecosystems, socioeconomic fabric, and culture and customs, geographic connotation is important. The government of India has also launched a major Geographical Information System (GIS) based database for planning such as National Urban Information Systems (NUIS) for major towns and cities and AMRUTH for more than 500 cities for developmental planning, where primary data sets are available for analysis in understanding the location of different crime hot spots and the processes therein.

At this juncture, this book, titled *Geographical Information System and Crime Mapping*, authored by Dr. Monika Kannan and Dr. Mehtab Singh, has aptly captured the basics of the GIS, its application, and different analytical methods it uses for identifying crime hot spots, inference as patterns and trends, and geo-visualizations of the results for better comprehension by students, researchers, planners, administrators, and executing agencies.

Professionally, I have known Dr. Monika Kannan and Dr. Mehtab Singh for the past decade. I keenly observed their enthusiasm for state-of-the-art technologies such as GIS and remote sensing applications. Their research acumen in exploring new technological tools in thematic applications is their specialization. This book is the outcome of their dedicated research and interaction with peers and users.

This book broadly covers an introduction to GIS; spatial analysis; spatial crime mapping; geography of neighborhood studies; identifying crime hot spots; and mapping for operational police activities, crime monitoring, and management. I have no doubt that this book will become a valuable reference volume for those who wish to engage themselves in GIS and its analysis for various themes and specifically for crime management. I heartily congratulate the authors for their painstaking effort in writing this elegant book, which will be immensely useful for all researchers, administrators, and students who wish to adopt digital methods of data processing using geographical connotation. This is only a beginning and I expect Drs. Kannan and Singh to author more books of this nature, which is the need of the hour. I wish them all the best in their future endeavors.

Y.V.N. Krishna Murthy
Formerly: Distinguished Scientist/ ISRO, *Director NRSC,*
Scientific Secretary ISRO, Director IIRS, Director CSSTEAP
(UN Affiliated Centre), Director NARL, Director ANTRIX Corporation,
Director RRSC's, Head RRSC Central

FOREWORD III

I am glad that Dr. Monika Kannan and Dr. Mehtab Singh have shared their experience in crime mapping with the help of geographical information science. Since crime is a major challenge that affects millions of lives around the globe, the regular or standard intelligence method of keeping criminal records is not sufficient to meet the current crime scenario requirements.

In the present scenario, when the number of crime incidences are increasing, manual processing of crime data is incapable of providing accurate, reliable, and comprehensive results to assist the decision support system. Conversely, GIS mapping plays an important role because of its powerful tool for visualization, analysis, and maintaining a huge database.

The authors have taken a very good initiative to associate both aspects. This book combines the topics of principles of crime occurrence, organizational structures for using crime mapping for crime analysts, planners, law enforcement agencies, and other stakeholders using GIS techniques, and provides a complete start-to-finish coverage of crime mapping.

The authors have several years of practical experience in the field of crime mapping and GIS which they have implemented in this book. I believe this book has great value to students, researchers, and decision makers.

I hope this book receives wide publicity and will be effectively used by the research community as it so richly deserves.

Prakash Chauhan, PhD
FNASc Director, IIRS, Dehradun

Preface

Security and safety challenges rank among the most pressing issues of modern times. Crime is one such challenge that impacts the lives of millions across the globe and features prominently in the public conscience and in governmental policies. The traditional and age-old prevailing system of intelligence and criminal record maintenance has lately been found inadequate to keep up with the requirements of the existing crime scenario. With the increasing number of crime incidences, manual processing of crime data seems to be incapable of providing accurate, reliable, and comprehensive data around the clock and does not help in trend prediction and decision support. In order to reduce the crime rate for the future, a comprehensive approach is required. Approximately 70% of crimes are committed by repeat offenders. If we are able to identify previous offenders and monitor them closely, then the crime rate would drop significantly. It is not as simple as it sounds. Identifying previous offenders is one of the toughest problems in countries such as India. There are information gaps everywhere, such as in the record keeping, digitization, access, accuracy, updates, and availability. To overcome this, there is a strong need for smarter planning, smarter investigations, and smarter responses. Hence, leveraging technological advancements is the way forward to ensure public safety and increase law enforcement effectiveness in crime prevention, control, tracking, and reduction. Analysts begin by aggregating crime data, identifying patterns and clusters, exploring the relationship between crime and other types of data sets, and assessing the effectiveness of crime-reduction strategies. All of these techniques rely on GIS mapping, which can play a major role because it is a powerful tool for visualization, analysis, and maintenance of a huge database. The coming years will see a convergence of technologies related to GIS navigation systems, advanced analytics, and predictive modeling using artificial intelligence, and the Internet, which will allow crime data to be analyzed in near real time and will help predict occurrence of criminal activities.

Delivered in an accessible style, this book combines the topics of theoretical principles of crime occurrence, GIS, analytical techniques, data processing solutions, information sharing, problem-solving approaches, map design, and organizational structures for using crime mapping for crime analysts, planners, educationists, law enforcement agencies, and other stakeholders. This book provides a complete start-to-finish coverage of crime mapping, including theory, scientific methodologies, analysis techniques, and design principles. It includes a comprehensive presentation of crime mapping applications for operational, tactical, and strategic management, and to demonstrate good practices.

This book is divided into seven chapters: The introductory chapter provides a generic introduction to the field of GIS and crime mapping that interests us most. In addition to explaining what GIS is and why it is such a powerful tool, this chapter covers such subtopics as development of the geographical information system and history of GIS development. It further elaborates on the essential theoretical concepts for interpretation and understanding of the geographical aspects of crime, which includes an explanation of how thinking about crime and space has developed into the field of geography and demonstrates many of the key theoretical concepts. Spatial analysis is the most intriguing and remarkable aspect of GIS, where we can combine information from many sources and derive new sets of information by applying a sophisticated set of spatial operators. These comprehensive collections of spatial analysis tools, some of which have also been explained in this chapter, extend our ability to answer complex spatial questions. The last section of the chapter discusses a wide range of applications, including GIS's role in crime identification, crime mapping, crime investigation, and display of spatial patterns of events and crime prediction.

While the mapping of crime remains important, especially with the growing sophistication of GIS, the subdiscipline has burgeoned to encompass a wider range of theoretical and empirical concerns. Chapter 2 presents and discusses a comprehensive range of theoretical and empirical concepts necessary for understanding and interpreting the geographical aspects of crime. It includes an explanation of how thinking about crime and space has developed into the field of spatial criminology. Three predominate theories— crime pattern theory, rational choice theory, and routine activity theory —have been discussed at length, with examples. To understand crime status and to measure the patterns of criminal activities, various methods such as location quotient, geographic profiling, crime and place, and crime hot spots, have also been discussed in this chapter.

On the basis of its strategic location, size, resources, demographic trends, and a rapidly expanding economy, India is considered as an emerging superpower of the world. However, India is facing various kinds of problems from within and across the border. Chapter 3 discusses insurgency, cross border terrorism, border disputes with neighboring countries, infiltration, and refugee movement, originating from across the border. Most of these problems are linked to various kinds of criminal activities in the country. Chapter 3 further reveals that India is among the top five countries in the world that is unsafe for women. An ever-increasing number of youth involved in heinous crimes is also making the country bleed internally and may have serious consequences for the country and society as a whole.

Crime mapping can play an important role in the crime reduction process, from the first stage of data collection to the monitoring and evaluation of any targeted response. It can also act as an important mechanism in a more pivotal preliminary stage, that of preventing crime by helping in the design of initiatives that are successful in tackling a crime problem. Chapter 4 discusses a wide range of techniques and tools to be used to detect and identify various types of crimes. Spatial interpolation analysis, using inverse distance weighting (IDW) and kriging, can be applied to highlight crime-prone zones and crime-susceptible areas, whereas the density tool can be used to create density surface to represent the distribution of crime incidents from a set of observations. On the other hand, hot spot analysis, using the spatial autocorrelation tool, helps to identify hot and cold spots in crime parlance. Regression analysis facilitates modeling, examining, and exploring spatial relationships, and helps crime analysts predict and observe crime patterns in a specific area. Most of these analyses use variables such as population density, occupational structure, poverty index, etc., for appropriate crime prevention or detection strategy.

To help law enforcement agencies curb crime in society, Chapter 5 focuses on how demography affects crime, discussing the various parameters such as the impacts of age, sex, and race, on deviant behavior, and the difference between compositional and contextual effects of demographic structure on aggregate crime rates. The intersection of criminal and demographic events in the life course and the influence of criminal victimization and aggregate crime rates on residential mobility, migration, and population redistribution, are some pertinent issues of analysis that have been discussed at length. Chapter 5 also highlights the role of network analysis for better understanding of the underlying causes of crime.

Law enforcement agencies need to include use of technology in their day-to-day mission of protecting citizens, ensuring officer safety, and serving the community. With a definitive GIS roadmap in place by all the law enforcement agencies, we can expect higher adoption of geospatial tools in all aspects of public safety and law enforcement, thus building a safer environment for all. Chapter 6 discusses various geospatial technique-driven gadgets that can immensely contribute in making law enforcement agencies robust and self-sustainable. The main focus of this chapter, however, is two case studies that may move the city police toward enhanced spatial crime mapping and demarcation of hot spot zones for better crime prevention strategies. One case study in Ajmer City (Rajasthan, India) illustrates, with examples, how usage of geospatial techniques can be handy in curbing crime occurrences in the study area. Another intensive micro-level study was conducted in the Dargah region of Ajmer City using geospatial techniques. Both of these studies revealed astonishing facts and relationships between urban growth, demographic profiling, and religious tourism with crime arousal, and its trends and patterns. The ability to collect and analyze data in real time brought GIS to a whole new level. This has been demonstrated by a safety application named "SATARK," which has been specifically devised for the women of Ajmer City. Putting it all in perspective, this chapter shows the myriad ways in which mapping technology using GIS can aid policing and crime reduction efforts.

Chapter 7 brings forth the various applications of implementing GIS in fields such as logistics operations leading to dramatic improvements in efficiency. This chapter also demonstrates the role of the safe city concept in crime management and monitoring and illustrating the policy frameworks and approaches to deal with the same.

This book is a result of several years of experience in applying GIS technology in crime mapping with practical concerns and real-world applications. Many of the terms and concepts introduced in this book have been compiled from a variety of different sources. These include some of the prevalent textbooks in the field, but most have been firmly implanted from practical experiences. This book is designed to be accessible, pragmatic, and concise.

Color figures are posted to book's webpage at https://www.routledge.com/9780367359034

Acknowledgments

First and foremost, we thank God. In the process of putting this book together, we realized how true this gift of writing is for us: "You have given the power to believe in our passion and pursue our dreams. We could never have done this without the faith we have in you, the Almighty."

Crime opportunities are neither uniformly nor randomly organized in space and time. Offering more than just theoretical or technical principles and concepts, the book applies GIS techniques to the real world; draws on global, regional, and local examples; and provides practical advice on mapping the built environment. This accessible text is essential reading for crime analysts as well as for all planners, urbanites, and geographers with an interest in how GIS can help us better understand the built environment from a socio-economic-cultural perspective.

Having an idea and turning it into a book is as hard as it sounds. The experience is both internally challenging and rewarding. We especially want to thank every individual who helped make this happen.

I, Dr. Monika Kannan, principal author, extend my indebtedness to the Indian Institute of Remote Sensing (IIRS), in Dehradun, India, for their insightful and incisive contributions in developing my interest in the subject and giving me an opportunity to contribute for the same.

I am deeply grateful toward my family for their unremitting support and love. My husband, Manoj, for always being there for me and being the comfort of my life; my two sons, Manan and Vihan, I cannot thank you all enough for adding luster and lending that extra sparkle to my life. You are the wind beneath my wings.

I am thankful to my co-author, Dr. Mehtab Singh, for his unstinting support and tireless efforts in bringing my thoughts to an explicit practice.

I, Dr. Mehtab Singh, co-author, owe an enormous debt of gratitude to the principal author of this book, Dr. Monika Kannan, who gave so freely of her time to discuss nuances of the text and pushed me to clarify concepts, explore particular facets of insight work, and explain the rationales for specific recommendations. This book represents our many years of collaboration; therefore, I cannot thank her enough for being my unrelenting source of inspiration to challenge how things get done in a corporate way. It's been a great privilege and honor to work with you, Dr. Kannan.

I want to thank my better half, Seema, for tolerating my incessant disappearances into my PC. A lifelong partner makes both the journey and destination worthwhile. Special thanks go to my son, Rishabh, and daughter, Twisha, for taking their precious time while I was writing this book.

We owe an enormous debt of gratitude to all our reviewers and expert assessors: Professor B.L. Sukhwal, Wisconsin University, United States; Professor Yukio Himiyama, Emeritus Professor, Hokkaido University of Education, Japan; Professor Y.V.N. Krishna Murthy, Registrar, IIST Kerala; Professor Shigeko Haruyama,

Mie University, Japan; and Professor Koichi Kimoto, Kwansai Gakuin University, Japan; for being the beacons of our life, leading and illuminating our paths along with dispelling the darkness of ignorance.

We express our sincere thanks to CRC Press, Taylor & Francis Group, for publishing the manuscript as a textbook as well as the Indian Society of Remote Sensing (ISRS) for considering our proposal for writing this book.

Monika Kannan and Mehtab Singh

Authors

Dr. Monika Kannan is a professor and head of the Department of Geography, Sophia Girls' College (Autonomous), Ajmer, Rajasthan, India. She has been engaged in teaching and research for the past 18 years, contributing specifically in the areas of crime mapping and analysis, geopolitics, geospatial technologies, gender issues, juvenile delinquencies, and border disputes. She has to her credit, seven books and nearly 60 research papers in journals and magazines of national and international repute. Dr. Kannan has received several awards such as the Certificate of Excellence by France Congress, Paris; Certificate of Merit by Rajasthan Public Service Commission, Govt. of Rajasthan, India; and National Young Geographer Award for her exemplary contributions in the field of research. Dr. Kannan is the chief editor of *Khoj–The International Peer Reviewed Journal of Geography*, along with being on the editorial boards of several other journals. She is the principal investigator for ICSSR Major Project (Impress Scheme) on Impact of Geographical Space and Urban Transformation on Women in Society: A Study of Ajmer City (Rajasthan). Dr. Kannan has been a consultant for the Institute of Development Study, Jaipur (Raj.) and Tata Institute of Social Sciences, Mumbai, for The Urban India Reform Facility, Ajmer and Pushkar's project. She has been certified by Harvard, Cambridge (USA), in data wise management techniques and has undertaken training from the prestigious Indian Institute of Remote Sensing, Dehradun, India, under the NNRMS program with specialization in geoinformatics.

Dr. Mehtab Singh is a professor in the Department of Geography, Maharshi Dayanand University, Rohtak, Haryana, India. He earned his doctorate from the University of Delhi. Dr. Singh has also undertaken training in the field of remote sensing and GIS from the prestigious Indian Institute of Remote Sensing, Dehradun, India. He has vast experience in developing content, as well as teaching and research experience in the field of geoinformatics. Dr. Singh has, to his credit, three books published by international and national publishers and more than a dozen research papers in national and international journals/books of repute. In addition to this, he has contributed to *Encyclopedia of Haryana*, a government of Haryana publication. Dr. Singh organized the prestigious International Geographical Union (IGU) Conference on Geoinformatics for Biodiversity and Climate Change in 2013. He has also organized many Capacity Building Training Programmes in geospatial technologies. He was principal investigator for the National Project on Development of the Framework for Networking Programme on Village Information System (VIS) funded by Department of Science and Technology, New Delhi. Dr. Singh was a member of the Task Force for Formulation of National Landslide Risk Management Strategy, NDMA, New Delhi. He is an accomplished speaker and he delivered keynote addresses in various national and international conferences held in India and abroad. Dr. Singh is also a member of various committees of UGC, Central and State Universities.

1 Introduction

The concept of geographical information systems (GISs) was subsequently researched and developed as a new discipline. Advancements in GIS are the result of several technologies. Databases, computer mapping, remote sensing, geography, mathematics, computer science, and computer-aided design all played a key role in the development of this discipline. The influx of the Internet saw widespread adoption of GIS heading into the millennium. GIS data has become more ubiquitous. Landsat, Sentinal, IRS, and even LiDAR data are accessible to download for free to users. Some of the online repositories store massive amounts of spatial data and now it's a matter of quality control and fitting it for needs. Gradually, the importance of spatial analysis for decision making has been recognized. Spatial analysis is the most intriguing and remarkable aspect of GIS where we can combine information from many sources and derive new sets of information by applying a sophisticated set of spatial operators. This comprehensive collection of spatial analysis tools extends our ability to answer complex spatial questions. The tools and techniques are now continuously refined to serve purposes such as urban planning, transportation, land use, land cover mapping, crime and disease mapping and analysis, utility services, resource planning and management, and many other applications. In addition, decisions that are made utilizing GIS and spatial analysis have important social implications for the people affected by them. For example, research into the geography of crime occupies something of a niche position and remains diverse and vibrant, cutting across many areas of social and cultural geography, as well drawing on and contributing to debates in criminology and other related disciplines. While the mapping of crime remains important, especially with the growing sophistication of GIS, the subdiscipline has burgeoned to encompass a wider range of theoretical and empirical concerns. Critical theories such as the Cartographic School, the Chicago School, the Factor Analysis School, and the Emergence of Environmental Criminology and Geography of Crime have been used to question the nature of crime and the extent to which it reflects wider inequalities in society.

1.1 DEVELOPMENT OF GEOGRAPHICAL INFORMATION SYSTEMS

GIS is more than just a software. People and methods are combined with geospatial software and tools, to enable spatial analysis, manage large data sets, and display information in a map and graphical form. The key word to this technology is *geography*. This means that some portion of the data is spatial. In other words, data that is in some way referenced to locations on the earth. As an emerging technology itself, the field of GIS is constantly evolving. Over the past few years, the global community has experienced revolutionary changes in this technology due to high-speed computer systems, variety of software and data, and answered geographic questions by designing and analyzing maps using user-selected criteria, thus leading to an extraordinary proliferation of applications. Over a span of 20 years, members of the geographic information community have seen this technology advance from command line, workstation-based software to tools that can now be used in the cloud and via mobile devices.

On the most basic level, geographical information systems technology is used as computer cartography, that is for straight-forward map making. The real power of GIS, however, is through using spatial and statistical methods to analyze geographic information. The end result of the analysis can be derivative information, interpolated information, or prioritized information into meaningful data used by individuals in endless possibilities. The information in GIS relates to the characteristics of geographic locations or areas. In other words, GIS answers questions about where things are or about what is located at a given site. The term "GIS" has different meanings in different contexts. It can relate to the overall system of hardware and software that is used to work with spatial information or designed to handle information about geographic features. It may also relate to an application; for example, a comprehensive geographic database of a country or region (Borneman, 2014).

Where is GIS headed in the future? This is probably the most-asked question posed to those in the GIS field and is probably the hardest to answer in a succinct and clear manner. While it can be difficult to try to prophesize the future of a technology, there are a wealth of clues that hint at and reveal a glimpse of the future of GIS. Geospatial data is everywhere. Customer addresses, time zones, office facility locations, service areas, political boundaries, status of shipments, utility networks, fieldworker positions, real estate, and location of mobile phones are all examples of geospatial data. Using GIS to leverage this information is critical for continued success. There are a variety of creative ways to use GIS to advance business and technology. The latest GIS technology developments make way for new applications and for the convergence of GIS with other emerging tools. The ability to collect and analyze data in real time brought GIS to a whole new level. Location-based services such as Uber are among the most common users for up-to-the-minute geographic information. Real-time geospatial intelligence can also be an effective way to track development during natural and man-made disasters. On the other hand, server-centric mobile applications help you to synchronize maps and data directly with a robust GIS server and cache information locally on the device to support both connected and disconnected applications. Home buying apps, real estate apps, and health and wellness apps are some of the mobile-based apps frequently used by consumers. Artificial intelligence (AI) is another rapidly evolving technology that has been applied to GIS for urban infrastructure, traffic congestion, air pollution monitoring, products, and services. The evolution of GIS is far from over, with exciting new software and methods continuing to change the way people utilize this technology.

1.1.1 History of GIS Development

The history of using geographic information to better understand and solve complex problems can be traced all the way back to the 19th century. The first documented application of what could be classified as a GIS was in France in 1832. French geographer Charles Picquet created a map-based representation of cholera epidemiology in Paris by representing the 48 districts of Paris with different halftone color gradients, essentially an early version of a heat map. The map, published in the report, *Rapport sur la marche et les effets du choléra-morbus dans Paris*, is likely the first use of spatial analysis in epidemiology.

Twenty-two years later, English physician John Snow took this concept a step further and demonstrated the problem-solving potential of maps by identifying the connection between an outbreak of cholera in London and a contaminated water supply. This was one of the earliest successful uses of a geographic methodology in epidemiology. While the basic elements of topography and theme existed previously in cartography, the John Snow map was unique, using cartographic methods not only to depict but also to analyze clusters of geographically dependent phenomena.

It also seems likely that the early stages of GIS development in the 20th century were characterized by individuals who were pursuing their goals in the field of GIS. Harvard Laboratory for Computer Graphics, the Canada Geographic Information System, the Environmental Systems Research Institute, and the Experimental Cartography Unit in the United Kingdom were the major players in the field during this period. There was no single direction until the field became the focus of intense commercial activity as satellite imaging technology meant that mass applications could be created for private use and the Environmental Systems Research Institute (ESRI) grabbed the opportunity with both hands.

The next significant step in the development of modern geographic information systems was in the early 20th century. A printing technique known as photo-zincography was used to separate out layers from a map. Vegetation, water, and developed land could all be printed as separate themes. While giving the appearance of being a GIS, this does not represent a full GIS as there is no opportunity to provide an analysis of the mapped data.

The first relevant experience that combined computers and geography can be found in 1959, when Waldo Tobler defined the principles of a system called MIMO (Map In–Map Out), with the purpose of applying computers to the field of cartography. He defined the basic ideas for creating, encoding, analyzing, and rendering geographical data within a computer system. Within the last five decades, GIS has evolved from a concept to a science. The phenomenal evolution of GIS from a rudimentary tool to a modern, powerful platform for understanding and planning our world is marked by several key milestones. There have been four distinct phases in the development of geographical information systems.

The early 1960s saw a new discipline being dominated by a few key individuals who were to shape the direction of future research and development. The two main reasons for this were the increasing need of geographical information and the appearance of the first computers. Initially, GIS was just a combination of ideas from quantitative cartography and the computer systems that existed at that time. It was basically the work of cartographers and geographers who tried to adapt their knowledge and their needs to a technology that looked promising. Since then, a large number of other disciplines have contributed to the field of GIS and their contributions are as important as those of cartography and geography.

Roger Tomlinson, widely acclaimed as the "father of GIS", during his time with the Canadian government in the 1960s, was responsible for the creation of the Canadian Geographic Information System (CGIS), which was an improvement over "computer mapping" applications, as it provided capabilities for overlay, measurement, and digitizing/scanning. In the mid-1960s, two applications, SYMAP and GRID, laid out the theoretical foundation for the analysis of raster and vector data, the two main approaches for encoding and storing geographical information. The main ideas for performing analysis in raster GIS were defined by Dana Tomlin with his map algebra. At the same time, society was becoming more concerned about the environment and the effect of human actions on it. This influenced GIS, which was becoming a fundamental tool for all tasks related to environmental management (land-use planning, environmental monitoring, resource planning and management, etc.), and boosted its development.

From the middle of the 1970s to the early 1980s, we saw the adoption of technologies by national agencies that led to a focus on the development of best practice. Once it was clear that GIS had a great future ahead, developments in spatial awareness and how to handle spatial data were being made in key academic centers such as Harvard. Conferences and symposiums were organized and GIS was already included in university curricula. Literature helped spread GIS to a wider audience. The industry of GIS consolidated itself and ESRI, the pioneer and most powerful leader of the GIS market, was founded during this period. First-vector GIS, called ODYSSEY GIS, was launched. By the late 1970s, two public domain GIS systems (MOSS and GRASS GIS) were in development.

Progress in computer memory and improved computer graphic capabilities saw the development and exploitation of the commercial market place surrounding GIS. By the early 1980s, M & S Computing (later Intergraph), Bentley Systems, ESRI, CARIS, MapInfo Corporation, and Earth Resource Data Analysis System (ERDAS) emerged as commercial vendors of GIS software. The first open-source GIS, GRASS (Geographic Resources Analysis Support System), appeared in 1985. This began the process of moving GIS from the research department into the business environment. By the end of the 1980s, cartography could be efficiently produced in personal computers, at a comparatively low cost, without the need of expensive and dedicated large mainframes.

Since the late 1980s, the focus has been on how to improve the usability of technology by making facilities more user centric. The adoption of GIS into the mainstream took off between 1990–2010. Processors are now on gigahertz. Graphic cards are crisper than they have ever been before. We now think of GIS data storage in terabytes rather than megabytes. Spatial and non-spatial data have become more ubiquitous. Optical (Landsat, IRS) and radar (LiDAR) data are accessible to download for free. The range of commercial GIS software products seems endless and digitized mapping data is more readily available. But what stands out is the major shift of GIS users building their own GIS software in an open, collaborative way. This type of software is made available to the public and is called open source. The big plus is that they are for public use at no cost. More light is shining on Quantum GIS (QGIS) than ever before. GIS and maps become even more important when linked to the huge data streams from the Internet of Things (IoT), allowing real-time monitoring of trends. The location intelligence derived from these streams of information when combined with artificial intelligence (AI), and predictive analytics can map out ways to drive productivity or adjust strategies to keep businesses on track to succeed.

Today, GIS has applications ranging from crime mapping and strategizing public health initiatives to selecting sites for archeological digs. With its movement to web and cloud computing, and integration with real-time information via the Internet, GIS has become a platform relevant to almost every human endeavor. As our world faces problems from expanding population, loss of natural resources, and widespread pollution, GIS will play an increasingly important role in how we understand and address these issues and provide a means for communicating solutions using the language of mapping. Over the history of GIS, researchers, programmers, and analysts have continued to innovate, developing fresh perspectives and technological breakthroughs. Their efforts resulted in the versatile tools and methods that empower projects for a wide variety of organizations.

1.2 GEOGRAPHY OF CRIME

Combining a geographical approach with an interest for crime research has significantly increased over the past few decades due to the fact that crime cannot be separated from the offender's habitat. This has resulted in development of human geography, a science describing and analyzing location patterns of static or moving phenomena of human origin on the surface of the earth. Crime phenomena are also a part of these phenomena, which means that they represent a subject of geographical research. There are no direct geographical theories that provide an explanation of spatial crime distribution, but a link between human geography and criminology has been established as a result of the development of a strong parallel that has existed in science for decades, similar to how criminology was predominantly put in the focus of sociology due to the series of paradigm shifts. The development period for spatial or environmental theories and empirical research of crime can be divided into the following phases:

1.2.1 The Cartographic School

By investigating environmental and spatial sources of crime, criminology as a scientific discipline has contributed to the development of the Classical School of criminal law in the late 18th century and in the beginning of the 19th century. That is how the Cartographic School (1830–1880) emerged in England and France as a part of the classical reformist school. Lambert Adolphe Jacques Quételet and Andre-Michel Guerre were the first to do detailed statistical studies of crime. Quételet found strong correlations between rates of crime and such factors as illiteracy, poverty, etc. He also noted that these variables remained the same as the highest crime rates continued to occur in the same parts of the city through several decades. The studies referred to the description of differences in attributes and the number of criminal acts between cities, regions, and smaller regions. Quételet (1831) and Guerry (1833) discovered that crime was unevenly distributed in different French districts. The study revealed that the majority of crimes against people and property occurred in the areas near the Seine,

Rhone, and Rhine Rivers and fewer crimes occurred in the central part of France. The proponents of the Cartographic School conducted studies of juvenile delinquency and professional crime, which showed that crime is a necessary expression of social conditions. According to Quételet and Guerre, crime is caused by the conflicts of values that arise when legal norms do not take into consideration the behavioral norms that are specific to various classes and interest groups living in certain geographic areas. The Cartographic School uncovered evidence that season, climate, sex, age, population composition, poverty, etc., were also related to criminality, most specifically that crime rates were greater in the summer among heterogeneous populations, and among the poor and uneducated they were highly influenced by drinking habits. This school identified many relationships between crime and social phenomena that still serves as a basis for criminal studies. As geography plays an important role, the Cartographic School can contribute valuable information to criminal research and crime prevention. One of the most important tools in identifying crime is crime mapping, which is mapping of crime using a geographical information system to conduct spatial analysis of crime problems.

1.2.2 The Chicago School

Scientists from the University of Chicago have studied and tried to provide an explanation of the distribution of crime in Chicago in the late 19th and early 20th centuries by introducing a concept of social ecology that consisted of two elements. The first element was based on social competition or social conflict due to scarcity of spatial resources of expanding cities caused by industrialization and urbanization. The second element referred to nature and quality of social organization within various areas. Thanks to the population boom, they were able to study in detail, over a short span of time, the shifts from inner city to suburbs, and the differences in crime rates between affluent suburbs and the inner-city poor (Treadwell, 2006). They were quick to draw a link between juvenile delinquency and the economic and geographic patterns of urban development. The most significant Chicago School researchers who applied the concentric zone theory were Shaw and McKay (1942), who established a link between delinquency in different neighborhoods and other socioeconomic factors. They found that certain areas had consistently high delinquency rates despite rapid turnover of the population, this tended to support the idea that the environment itself was at least partly responsible for generating crime. According to Treadwell (2006), the idea of social ecology holds that crime is a response to unstable environment and abnormal living conditions. To establish this link, Shaw and McKay identified what became known as "white flight"—the phenomenon of well-off, well-educated people moving out of urban centers to more affluent suburbs, leaving cities with concentrations of poor, less-educated citizens, often concentrated in ethnic or racial groups. This pattern of movement and separation helps explain the observation that certain areas are more crime-prone. It is not the result of more criminals flocking to certain areas, but rather that the bad living conditions and poor infrastructure create barriers in community, and offer opportunities or incentives for criminal behavior. This is based on the idea that if people are concentrated in areas with limited opportunity and/or close proximity to criminals, they are more likely to learn deviant behavior. Theories developed by the Chicago School are still central tenets of criminology, whether modern researchers agree or are trying to discredit them. For almost a century, the Chicago School has held its place in criminology, but as society changes and its needs change this long tradition could also be displaced.

1.2.3 The Factor Analysis School

Empirical research conducted in the mid-20th century had many things in common with the Chicago School. It highlighted a link between the geographical distribution of offenders and criminal acts and characteristics of certain areas inhabited by delinquents (i.e., places where offenses are committed). The studies done by Morris and especially Lander in relation to Shaw and McKay's research results did not show substantial differences in crime analysis of large-city agglomerations.

1.2.4 The Emergence of Environmental Criminology and Geography of Crime

During the 19th century and into the early 20th century, crime was considered a socially derogated, deviant form of behavior that is different from the usual, frequent behavior patterns, and the subsequent studies were focused on the origin of crime motivation (Ksenija and Jelena, 2017). On the last page of his seminal book, *Crime Prevention Through Environmental Design*, Jeffery (1971) coined the term "environmental criminology" in a call for the establishment of a new school of thought in the field of criminology. The focus of the new school was the environment within which crime occurs, not the individual offender. However, the main principle of the Classical School of criminology (i.e., the deterrence of crime before it occurs) still held the prime focus. The idea that Jeffery put forth was to make crime a high-risk and low-reward activity; create environmental contingencies that control land use, travel paths, and access; and, in the long term, create a society in which the existing laws are respected, potential offenders are busy in jobs and/or education, citizens are given the knowledge to protect themselves through neighborhood organization and individual actions (Andresen, 2009).

Newman's (1972) book, *Defensible Space: Crime Prevention Through Urban Design*, propounded that we need to build our neighborhoods in such a manner that fosters the development of a social cohesion that acts against crime, using defensible space. According to Newman, defensible space is a model of environments that inhibit crime through the creation of the physical expression of a social fabric that defends itself. This environment is dominantly created through changes in architecture. The works of Jeffery and Newman have strongly influenced the research of physical space and environment within which criminal acts occur. One such environmental criminology approach was developed in the 1980s by Paul and Patricia Brantingham, putting focus of criminological study on environmental or context factors that can influence criminal activity. These include space, time, law, offender, and target or victim. These five components are a necessary and sufficient condition, for without one, the other four, even together, will not constitute a criminal incident (Brantingham and Brantingham, 1991). Theories within social ecology focus on places where individuals commit their crimes. The offenders do not choose their targets by chance, but on the basis of awareness space and a series of rational choices they make, given the fact that a criminal commits a criminal act when highest gain is likely to occur and the risk of getting caught is at its lowest. Routine activity theory is an attempt to identify, at a macro level, criminal activities and their patterns through explanation of changes in crime rate trends (Cohen and Felson, 1979). Generally speaking, all the regular activities that we undertake to maintain ourselves (work, school, shopping, recreation) are our routine activities. Routine activity theory is a good example of an integration of approaches such as victim, time, and place. The offender does not choose a place for committing a crime randomly, but chooses a location within his awareness space. This is a space that is familiar or known to him because he is informed about it and feels safe in it. Generally, awareness space encompasses a neighborhood where a person lives, place and surroundings in which that person works or acts in any way, and roads that connect all this and a workflow of activities that this person performs in this space.

Though environmental criminology is a part of criminology but with the development of behavioral approach in the geography of crime, both of the disciplines tend to merge with each other. As crime perspective has changed radically during the last few decades, geography of crime has become a distinct discipline of geography, with the emergence of environmental criminology. According to Georges (1978), "The geography of crime is the study of the spatial manifestation of criminal acts. It is the study of the social and cultural organization of criminal behavior from a spatial perspective. It is not the study of spatial organization and particularities of the judicial response of criminal behavior."

1.3 UNDERSTANDING SPATIAL ANALYSIS

Spatial analysis encompasses everything we do with our geospatial data, from framing our research question to presenting our final results. It allows us to solve complex location-oriented problems and better understand where and what is occurring in a particular geographic space. In the present, spatial analysis goes beyond mere mapping to study the characteristics of places and the relationships between them, rather it involves using analytical techniques to examine geospatial data and answer questions by highlighting or creating new information. Thus, spatial analysis lends new perspectives to decision making in a number of ways. Whenever we look at a map, we inherently start turning that map into information by analyzing its contents and finding patterns, assessing trends, and making decisions. Thus, spatial analysis refers to statistical analysis based on patterns and underlying processes. It is a kind of geographical analysis that elucidates patterns of personal characteristics and spatial appearance in terms of geostatistics and geometrics, which is known as location analysis. The entire process involves statistical and manipulation techniques, which could be attributed to a specific geographic database (Cucala et al., 2018).

Classification of the techniques of spatial analysis is difficult because of the many different fields of research involved. They have been invented in many disciplines, including mathematics, geometry; statistics, spatial statistics and statistical geometry; and in geography and other earth sciences. Spatial analysis is a type of geographical analysis that explains the behavioral patterns of humans, animals, epidemics, etc., and their spatial expression in terms of geometry. Thus, the range of methods deployed for spatial analysis varies with respect to the type of the data model used. Measurement of length, perimeter, and area of the features is a very common requirement in spatial analysis (Clark and Evans, 1954). However, different methods are used to make measurements based on the type of data used, i.e., vector or raster. Invariably, the measurements will not be exact, as digitized features on a map may not be entirely similar to the features on the ground, and moreover, in the case of raster, the features are approximated using a grid cell representation (Oliver and Webster, 2007).

Spatial analysis is one of the most intriguing and remarkable aspects of GIS. Due to the flexibility of GIS, spatial analysis can constitute one simple task or a series of complex tasks. A simple spatial analysis process might consist solely of visualizing data on a map for users to interpret, and a complex spatial analysis process can incorporate multiple datasets, spatial statistics, and Python scripts. These tools enable you to address critically important questions and decisions that are beyond the scope of simple visual analysis. However, complex issues arise in spatial analysis, many of which are neither clearly defined nor completely resolved, but form the basis for current research.

1.3.1 Types of Spatial Analysis

1.3.1.1 Autocorrelation

In order to detect the spatial pattern (spatial association and spatial autocorrelation), some standard global and new local spatial statistics have been developed. These include Moran's *I*, Geary's *C*, and G statistics (Getis et al., 1992). All of these spatial analytical techniques have two aspects in common. First, they start from the assumption of a spatially random distribution of data. Second, the spatial pattern, spatial structure, and form of spatial dependence are typically derived from the data (Bao, 1999). Spatial autocorrelation is simply looking at how well objects correlate with other nearby objects across a spatial area. Positive autocorrelation occurs when many similar values are located near each other and vice versa in cases of negative autocorrelation. The importance of spatial autocorrelation is that it helps to define how important spatial characteristics affect a given object in space if there is a clear relationship (i.e., dependency) of objects with spatial properties. Strongly positive or negative results indicate that a clear spatial property is found in the object with a high correlation (Griffith, 2011). Perhaps the most common way in which autocorrelation is measured is using Moran's *I*, which allows the correlation measure to determine correlation based on multiple dimensions across a given space. Results are generally used to measure how well an object correlates globally, that is across a given defined space for a dataset. Geary's ratio *C* is another similar measure, where this measure is more sensitive to local variations and can be used to define local patterning within a dataset (O'Sullivan and Unwin, 2010).

1.3.1.2 Spatial Interpolation

Spatial interpolation methods estimate the variables at unobserved locations in geospace based on the values at observed locations because it is often difficult to obtain height, magnitude, or concentration of a phenomenon from every location of a given area. Instead, one can measure the phenomenon at strategically dispersed sample locations (randomly or regularly spaced) and create a continuous surface by predicting values for all other locations. Various interpolation techniques such as Inverse Distance Weighted (IDW), Sapline and kriging interpolation, as well as polynomial trend and natural neighbor methods can be used to estimate rainfall, elevation, temperature, or other spatially continuous phenomena. IWD and Spline methods are deterministic interpolation methods because they assign values to locations based on the surrounding measured values whereas kriging is based on statistical models that include autocorrelation, the statistical relationship among the measured points. Analysts can also create non-traditional surfaces using various other functions (i.e., density surface showing the density of objects), such as persons per square kilometer of land area; distance-based surfaces showing distance to various features, such as a police station, petrol pump, bank, city market, etc. Using surface analysis in GIS, we can obtain elevation from a terrain surface, or density areas for crime analysis. These techniques not only have the capability to produce a prediction surface but also provide some measure of the certainty or accuracy of the predictions.

1.3.1.3 Spatial Regression

Spatial regression method capture spatial dependency in regression analysis to avoid statistical problems such as unstable parameters and unreliable significance tests. It also helps to provide information on spatial relationships among the variables involved. Depending on the specific technique, spatial dependency can enter the regression model as relationships between the independent variables and the dependent one. Geographically Weighted Regression (GWR) is a local version of spatial regression which allows to study how a given phenomenon varies spatially in a particular area (Fotheringham et al., 2002). For example, in crime-related studies, spatial regression method can be applied successfully to understand what variables (education, occupation, age, income, etc.) explain crime locations, which can be used for decision making. Spatial regression model can be used to predict future crime locations.

1.3.1.4 Spatial Interaction

Much of the data gathered today contains at least one location component. Often times, there will be multiple location components or additional factors within a single data set. In such cases, there are many valuable questions to ask about how these location components or factors might impact one another. For this, one needs to draw on specific spatial techniques to plumb the depths of available data for answers to these questions. Various forms of spatial interaction models have been applied in aggregate analysis, most commonly the spatial interaction or gravity model. The gravity models provide a flexible approach for the analysis of spatial interactions between spatially separated nodes, being applied in a wide variety of studies, such as those devoted to migration, commodity flows, traffic flows, residence-workplace movement, market area boundaries, etc. In general terms, the gravity models state that the interaction between two centers is in direct proportion to their size and in inverse proportion to the distance between them. After specifying the functional forms of this interaction, the expert can estimate model parameters using observed flow data and standard estimation techniques such as ordinary least squares or maximum likelihood. Competing destinations versions of spatial interaction models include the proximity among the destinations in addition to the origin-destination proximity. This captures the effects of destination (origin) clustering on flows. Artificial Neural Networks (ANN) can also estimate spatial interaction relationships among locations and can handle qualitative data.

1.3.1.5 Simulation and Modeling

In the present day, our ability to collect and organize observations and to combine and transform data to generate new information, including maps and models, make GISs an essential tool for designers and policy analysts. The observations (data) might be used together to understand real and hypothetical situations. Geographic models may be useful for developing and communicating an understanding of how things and conditions affect each other. This is the sort of information that designers and policy analysts often have to make. Geographical models can be useful for conducting hypothetical (what-if?) experiments that explore plausible ways that critical aspects of situations might be affected by change. Spatial interaction models are aggregate and top-down: they specify an overall governing relationship for flow between locations. This characteristic is also shared by urban models such as those based on mathematical programming. Important spatial simulation methods are cellular automata and agent-based modeling. Cellular automata modeling imposes a fixed spatial framework such as grid cells and specifies rules that dictate the state of a cell based on the states of its neighboring cells whereas Agent-based modeling uses software entities that have purposeful behavior and can react, interact and modify their environment while seeking their objectives. While cellular automata are of interest in spatial modeling and often used to model land cover changes, whereas agent-based models are being applied for many operations including managing traffic flow. Cellular automata and agent-based modeling are divergent yet complementary modeling strategies. They can be integrated into a common geographic automata system where some agents are fixed while others are mobile.

1.3.2 Data Types in Spatial Analysis

According to Camara et al. (2004) the most used taxonomy to characterize the problems of spatial analysis consider three types of data:

Events or Point Patterns – phenomena expressed through occurrences identified as points in space, denominated point processes. Some examples are crime spots, disease occurrences, and the localization of vegetal species.

Continuous Surfaces – estimated from a set of field samples that can be regularly or irregularly distributed. Usually, this type of data results from natural resources survey, which includes geological, topographical, ecological, phitogeographic, and pedological maps.

Areas with Counts and Aggregated Rates – means data associated to population surveys, such as census and health statistics, and that are originally referred to individuals situated in specific points in space. For confidentiality reasons, these data are aggregated in analysis units, usually delimited by closed polygons (census tracts, postal addressing zones, municipalities).

The previously mentioned data types are environmental and socioeconomic in nature, which requires solution through a set of chained procedures that aims at choosing an inferential model that considers the spatial relationships present in the phenomenon and their dependency patterns. In the case of point pattern analysis, the objective is to study the spatial distribution of the points under consideration, testing hypothesis about the observed pattern, if it is random or is regularly distributed (Table 1.1). For surface analysis, the objective is to reconstruct the surface from which the samples were removed and measured. For a real analysis, the areas are usually delimited by polygons with internal homogeneity, that is, important changes only occur in the limits.

1.3.3 Geovisualization

Geovisualization combines scientific visualization with digital cartography to support the exploration and analysis of geospatial data and information, including the results of spatial analysis. Geovisualization is generally defined as the method of interactively visualizing geographic information in any of the spatial analysis measures, although it may also apply to the visual production (e.g., charts, maps, diagrams, 3D views) or the techniques associated with it. In contrast with traditional cartography, this method is typically three- or four-dimensional and user-interactive. GISs increasingly provide a range of such tools, providing static or rotating views, draping images over 2D surface representations, providing animations and fly-throughs, dynamic linking, and spatio-temporal visualisations. A key argument for geovisualization is that visual thinking using maps is central to the creation of scientific processes and theories, and the role of maps has expanded beyond communicating the end results of an experiment or documentation. As such, geovisualization integrates with a variety of disciplines including cartography, visual analysis, knowledge visualization, scientific visualization, statistics, computer science, art-and-design, and cognitive science.

1.4 GIS AND ITS APPLICATIONS

1.4.1 Urban

Cities are systems of tremendous, ever-evolving complexities. Guiding a city development requires spatial information that's robust, nuanced, and constantly updated, as well as the problem-solving skills to apply that information. Spatial data points the way to improving quality of life and building sustainable communities, while GIS professionals use spatial thinking to transform that data into actionable insight and solutions. For urban planning, GIS is being used as an analytical and modeling tool, which is needed to design and map the city landscape. These tools can be applied to a wide array of problems, which comprises addressing problems related to data base structures, simple and complex analytical models alike. Visualization, spatial analysis, and spatial modeling are the most frequently used GIS functions in plan making. GIS can help to store, manipulate, and analyze physical, social, and economic data of a city. Planners can then use the spatial query and mapping functions of GIS to analyze the existing situation in the city.

TABLE 1.1
Important Operations of Spatial Analysis in GIS Environment

Operation	Description
Measurement, basic geoprocessing	• Location, distance, area; point in polygon; line on polygon overlay; polygon overlay; nearest neighbor search; buffering, merge, dissolve, clip, intersection, union etc.; spatial join; raster processing (classification, logical/airthmatic operations, aggregation); vector-raster conversion
Proximity and contiguity analysis	• Proximity and contiguity analyses are respectively simple methods of determining and indicating measures of distance between locations, or of showing a location's degree of adjacency to neighboring locations.
Intervisibility	• Defines, from map evidence, whether or not it is possible to have a direct line of sight between any two points on the map. Thus a calculation is made, bearing in mind the existence of high ground as shown by contours, whether or not hills or other high ground would obscure the line of vision.
Network analysis	• Simple forms of network analysis are covered in shortest route and connectivity. More complex analyses are frequently carried out on network data by electrical and gas utilities, communications companies etc. These include the simulation of flows in complex networks, load balancing in electrical distribution, traffic analysis, and computation of pressure loss in gas pipes.
Trend surface analysis	• Method for establishing whether a generalized spatial surface exists, (i.e., one which may be obscured by a mass of detail in the real world). For instance, in any one country there may be an overall "wealth" surface which trends perhaps from east to west but which could well be obscured by numerous pockets of prosperity or poverty. From the marine viewpoint, it is quite likely that trend surfaces would exist with regard to the distribution of particular species (i.e., such that they would gradually decline outwards from a biologically optimum area but in an irregular, and thus perhaps obscured, way).
Location optimization	• Widely used as a GIS-based method which allows for the selection of optimum locations for the siting of any activity. This analysis is usually used by larger commercial companies when seeking, for instance, sites for new retailing outlets or for centralized distribution points. Similar analyses are also used by the forestry and agricultural sectors in seeking to optimize their operations, though here physical rather than economic criteria might be more important.
Digital terrain modeling	• Process whereby it is possible, using digitized height data, to build a 3D model of any desired area. These models are also be called 2.5D since they only show surface heights and not true volumetric data.
Complex correlation analysis	• The ability to compare maps representing different time periods, extracting differences or computing indices of change.

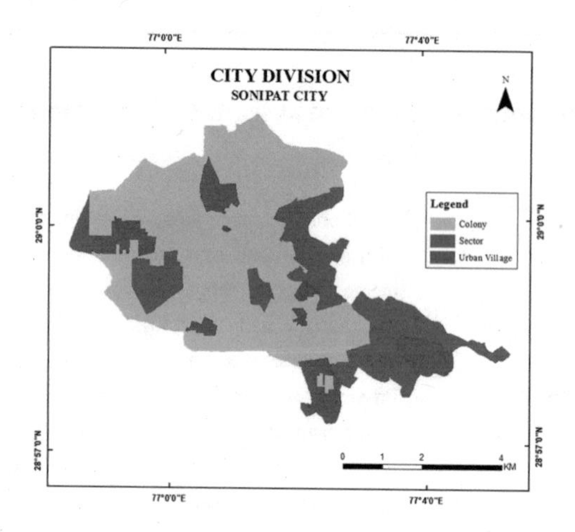

FIGURE 1.1 GIS Application in Urban Studies.

Fig. 1.1 depicts application of GIS in urban studies. Using this information, planners can use overlay analysis, where GIS can help to review and analyze checks on regulatory compliance, review of environmental impact, preservation of historic sites, regional planning beyond the borders of a city or town, and mapping the delivery of utilities and planning for service interruptions. The applications of GIS in urban planning, especially in areas of spatial modeling, have improved manyfold in the past two decades. 3D printing, water main breaks, participatory GIS, toponymy, estimating run-off and sources of storm water pollution, Integrating urban development visions with GIS such as smart urban planning, smart utilities, smart transportation, and smart public works are some of the examples.

1.4.2 Land Use

Making decisions for rational, sustainable land use is becoming increasingly complex as land pressure and the competition for land, and extent of land degradation problems increase. The information and knowledge required for these decisions should be based on comprehensive and quantified assessments of potentials and development possibilities of the land resources, taking into account the biophysical, environmental, socioeconomic factors, as well as the space and time dimensions of sustained land use. For a sustainable land use plan, nowadays, land use planning requires more and more data integration, multi-disciplinary and complex analysis, and needs faster or more precise information. Certainly, GIS, which has a strong capacity in data integration and analysis and visualization, become the main tool to support land use planning approaches. A land use map prepared in a GIS environment is depicted in Fig. 1.2. The outputs from such assessments are required by land use planners, ecologists, economists, environmentalists, researchers, agronomists, and other land users, corresponding to various areas of applications such as land suitability and land productivity assessment, participatory land use planning, land

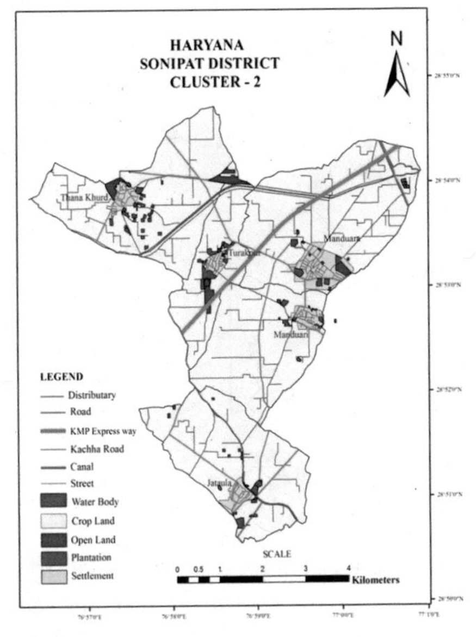

FIGURE 1.2 GIS Application in Land Use Classification.

degradation assessment, urban sprawl estimation, assessment of urban heat islands, network analysis, population estimation and distribution, flooding assessment, land use/land cover change detection, land evaluation, urban model development feasibility, Land-use Conflict Identification (LUCIS) model, Cloud computing[en dash]based land base mapping, open street map, 3D viewshed, space syntax models, quantification of land resources constraints, optimal resources planning, land management, agricultural technology transfer, agricultural inputs recommendations, farming systems analysis and development, environmental impact assessment, monitoring land resources development, agro-ecological characterization for research planning, agro-economic zoning for land development and nature conservation, and ecosystem research and management.

1.4.3 Transportation

The application of GIS has relevance to transportation due to the spatially distributed nature of transportation related data, and the need for various types of network level analysis, statistical analysis, spatial analysis, and manipulation. Its graphical display capabilities allow not only visualization of the different routes but also the sequence in which they are built, which allows the understanding of the logic behind the routing network design. At a GIS platform, the transport network database is generally extended by integrating many sets of its attribute and spatial data through its linear referencing system. In addition to this, GIS will facilitate integration of all other socioeconomic data with transport network database for wide variety of planning functions. The use of GIS for transportation applications is widespread (Fig. 1.3). Typical applications include highway maintenance, traffic modelling,

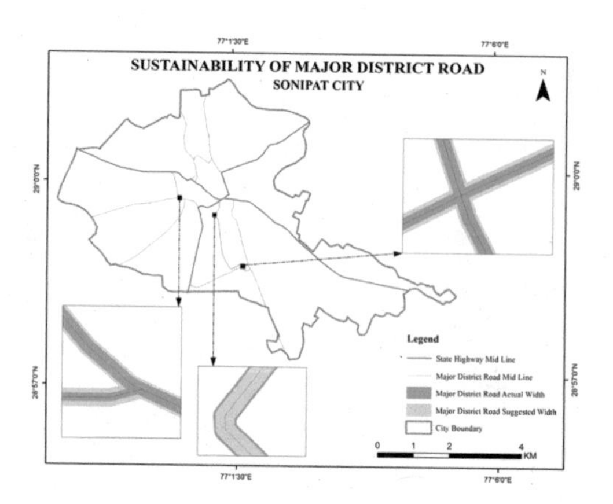

FIGURE 1.3 GIS Application in Transportation Network Studies.

accident analysis, route planning and environmental impact assessment of road schemes, whereas potential applications for GIS in transportation planning and management system are Transportation System Management (TSM), travel demand forecasting, pavement management, traffic engineering, planning and research, bridge maintenance, road safety management, corridor preservation and right-of-way, construction management, hazardous cargo routing, overweight/oversize vehicles permit routing, and accident analysis. Other planning applications include evacuation planning, planning for hazardous material release incidents, development of new traffic analysis zones from census tracts, urban traffic air pollution, traffic congestion, shortest path, Intelligent Transportation System (ITS), and development of new urban highway networks. The interaction between the transportation system and its surrounding environment makes the GIS technology ideally suited for transporting hazardous material, routing design, risk analysis, and decision making. GIS can also be integrated with sophisticated mathematical models and search procedures to analyze different management options and policies.

1.4.4 Environment

In the process of human evolution, the issues confronting today are safe guarding the natural environment and maintaining good quality of life. While taking up developmental activities, the assimilative capacities of the environmental components (i.e., air, water, and land) to various pollutions are rarely considered and because of the overuse, congestion, and incompatible land use, environmental pollution, land degradation, etc. are becoming heated topics in environmental studies. In this scenario, GIS can play a vital role for analysis and in formulating the quick mitigation plans for high-risk environments (Fig. 1.4). GIS allows better viewing and understanding of physical features and the relationships that influence in a given critical environmental condition. GIS supports activities in environmental assessment, monitoring, mitigation and can also be used for generating environmental models. Apart from data analysis, GIS can also help

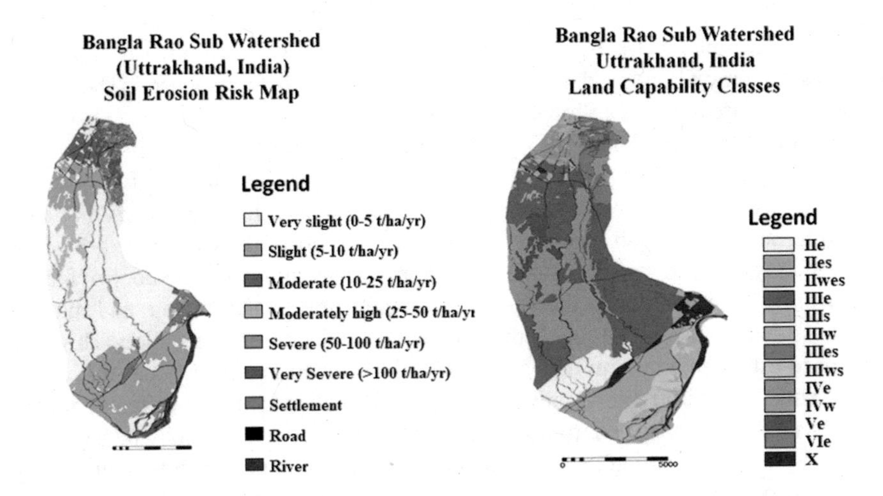

FIGURE 1.4 GIS Application in Environmental Studies.

the environmental data analysts in the field because most of the GIS tools are flexible enough to work in the field to give the exact location of damage and amount of devastation. GIS is capable of providing solutions in the areas of managing natural resources, wild land analysis, soil mapping, waste water management, air pollution and control, disaster management, zoning of landslide hazards, estimation of flood damage, forest fires management, sea water and fresh water interface studies, hazard mitigation and future planning, oil spills and its remedial actions, emergency services such as fire prevention, forest fires management, identification of volcanic hazards, and coal mine fires. Other GIS applications include wetland inventory, invasive species modular dispersal, dead zones, site remediation, etc.

1.4.5 Crime

Crime is a human phenomenon; therefore, its distribution in space is not uniform. Over the last few decades, a new worldwide socioeconomic order led to an increasing number on crime rates. The traditional and age-old prevailing system of intelligence and criminal record maintenance has failed to live up to the requirements of the existing crime scenario. Manual processes neither provides accurate, reliable, and comprehensive data around the clock nor does it help in trend prediction and decision support. To overcome this, there is a strong need of smarter planning, smarter investigations, and smarter responses. Hence, a need is felt for the effective use of information technology for crime prevention, control, tracking, and reduction. In this situation, GIS can play a major role because it is a powerful tool for visualization, analysis, and maintaining a huge database. GIS is such a powerful technology that has the capability to create a single visual output that combines multiple data layers into a meaningful output. The capability of GIS is high in different spatial units, providing immediate results and linking the spatial and other class of information for better crime analysis. It also helps to trace out the event causes, thereby allowing decision makers with immediate responses. The various application areas are as follows:

Crime Identification: GIS offers a broad platform for determining location coordinates. Normally, police records are fundamental instrument of data collection. However, in many of the advanced countries, location technology is widely used throughout the call reception process and as part of the call-routing function, where the emergency call is routed to the appropriate command and control facility. GIS provides capability to the agencies to better understand, update and maintain address databases, crucial for identifying and providing precise location. This helps them to establish a system that geo-codes addresses to tackle criminality.

Crime Mapping: Crime prevention can be looked at from two different angeles: firstly averting any persons from starting to commit crime and secondly intervening in persons who have committed offences before any additional activity results. Crime mapping offers considerable scope to intercede in both of these situations. Maps offer to crime analysts the graphic representations of crime-related issues. Mapping enables both specialists and non-experts to picture and analyze crime hot spots so that an understanding of where and why crimes occur can improve attempts to combat crime. Mapping crime can help police protect citizens more effectively. Simple maps that display the locations where crimes or concentrations of crimes have occurred can be used to help direct patrols to places they are most needed. Policy makers may also use these maps to observe trends in criminal activity.

Crime Investigation: Information management has always been a main concern for law enforcement agencies, especially the location information. Traditional approaches, such as confidential information through informers, street investigations, etc., are effective ways of data collection. However, nowadays, the GIS technique is used to create geographical profiling of offenders, an investigative method that allows police to identify locations of connected crimes to help determine where an offender may live, particularly in serial cases. Moreover, crime scene investigators rely more on GIS techniques to document crime scenes using easy-to-configure web applications. Locations can be collected using high accuracy navigation devices. A GIS can be linked to satellites that capture live images to track moving suspects who might be escaping from crime scenes. Adding content such as recently flown drone imagery or 3D data makes maps more powerful tools for accident re-creation, investigation, or for courtroom presentations.

Display Spatial Patterns of Events: Over the last few years, crime analysis has become a general term that includes a lot of research subcategories: intelligence analysis, criminal investigative analysis, tactical crime analysis, strategic crime analysis, operation analysis, administrative crime analysis, etc., which may require a lot of data mining. GIS and remote sensing enable different combinations of available information and data to be used to identify the patterns of the crime by analyzing a range of variables recorded. Analysis and mapping of these patterns can be comfortable to see the relationship between crime, community characteristics, weather, and geographic features. The ability to unmask these crime patterns is very helpful as it can be used by security agencies to mitigate crime accordingly. It also helps to trace out the event cause, thereby allowing decision makers immediate responses.

Crime Prediction: After pattern analysis is done, crime prediction can now be simple to realize. Past trends are a foundation of what might in the future. Analytical techniques, particularly quantitative techniques, are applied to identify likely targets for police intervention and prevent crime or solve past crimes. A step further ahead, a mobile phone location identification in a cellular network based on the signal strength and angle of radio waves with a cell phone has become boon for crime prediction and identification for law enforcement agencies. Nowadays, web-based applications use open data on crime and combine spatial and temporal information to determine a crime risk index for given areas and times. This index helps determine the likelihood one could be affected by crime using the historical pattern through an entropy weighting procedure.

1.4.6 Land Information System

A land information system is a geographical information system for cadastral and land-use mapping, used by governments agencies. It consists of an accurate, current, and reliable land records in the form of cadastral information and its associated attribute and spatial data that represent the legal boundaries of land tenure and provides a vital base layer as a stand-alone solution that allows data stewards to retrieve, create, update, store, view, analyze, and publish land information (Fig. 1.5). The application of GIS technology has helped provide the necessary data sets of various land information to facilitate the work of planners. The immediate advantage is seen in the handling of various planning applications for development and in the formulation of master plans, structure plans, and local plans.

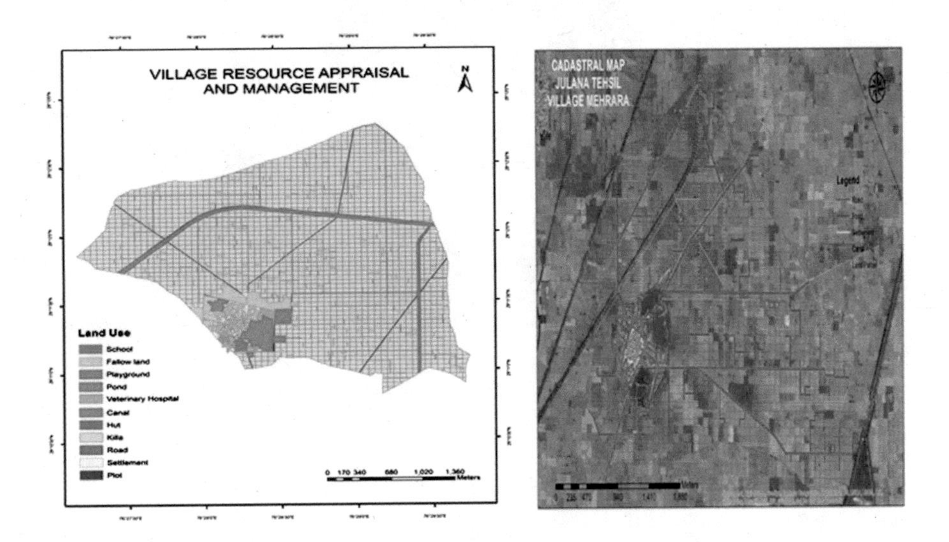

FIGURE 1.5 GIS Application for Land Information System.

1.4.7 Agriculture

As per the Food and Agriculture Organization (FAO) of the United Nations, the world's population will reach 9.1 billion, 34% higher than today's population, by 2050. Due to this expected growth, there is pressure throughout the world for higher agricultural production and reliable crop status information. To achieve sustainability in agricultural production, judicious use of natural resources (soil, water, livestock, plant genetic, fisheries, forest, climate, rainfall, and topography) in an acceptable technology management under the prevailing socioeconomic infrastructure is required. While natural inputs in farming cannot be controlled, they can be better understood and managed with GIS applications (Fig. 1.6). Farming is getting smarter with the availability of advanced technologies such as precision equipment, the Internet of Things (IoT), sensors and actuators, geo-positioning systems, big data, unmanned aerial vehicles, robotics, etc. Thus, GIS can substantially help in effective crop yield estimates, soil amendment analyses, and

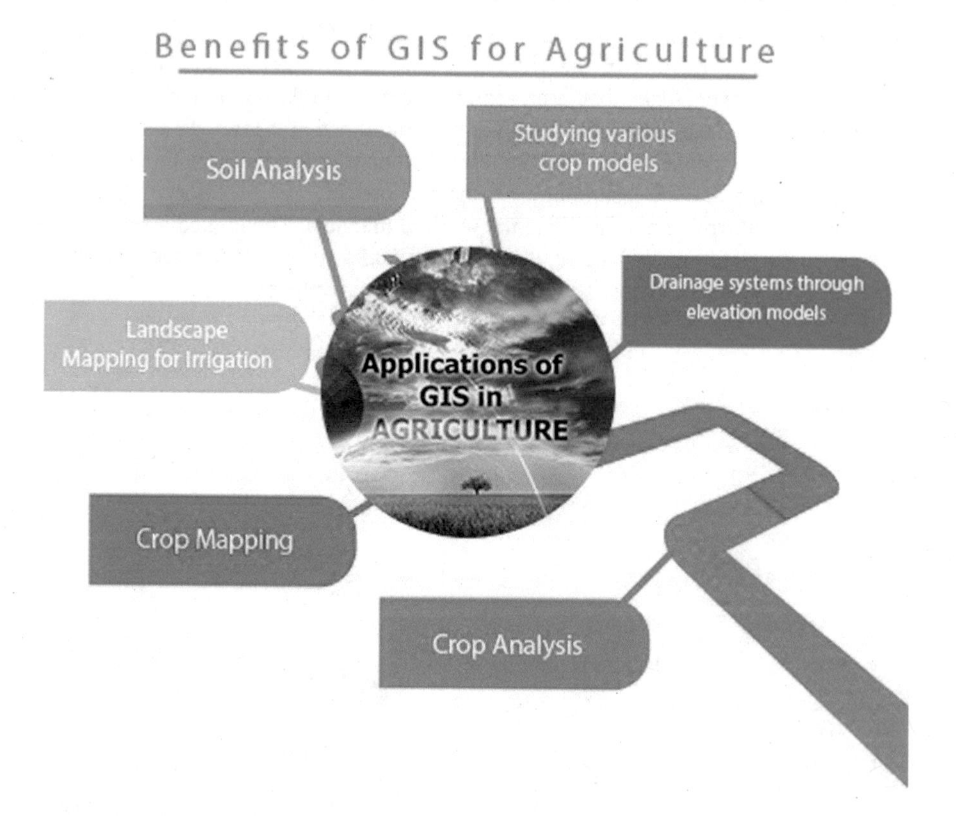

FIGURE 1.6 GIS Application in Agricultural Studies.

erosion identification and remediation. More accurate and reliable crop estimates help reduce uncertainty. GIS tools and online web resources can help farmers to conduct crop forecasting and manage their agriculture production by utilizing multispectral imagery collected by satellites, fixed-wing aircraft or unmanned aerial vehicles (UAVs), and processed to provide NDVI and other vegetation soil indices, to identify crop stress, disease control, real-time crop yield, crop assimilation model, crop resilience to climate change, crop productivity, agriculture capability, and crop forecasting are many other GIS applications for sustainable agriculture production. The ability of GIS to analyze and visualize agricultural environments and workflows has proven to be very beneficial to those involved in the farming industry.

1.4.8 Forestry

Forests have always been considered as the most important renewable resources and have a significant role in preserving the environment. Replacing conventional methods, GIS is important technology that is being widely used in public policy-making for forest and environmental planning and decision making. It is a good tool as it answers the questions that helps foresters in forest management activates such as location, condition, trends, patterns, and modeling (Fig. 1.7). Apart from that the major areas where use of GIS can be really useful in terms of forest management are in forest inventory mapping, biomass estimation of forest, canopy extraction, harvest estimation and planning, diversion of territory with respect to encroachment of human activity, forest fire mapping, river mapping for proper water resource to the forest environment, identification of deforestation, wildlife habitat conservation forest rehabilitation, conservation and biodiversity, forest carbon reserves (sequestering carbon), agent-based simulation, vertical point profile, global forest watch, leaf area index, 4D GIS, wildfire simulation, climate change, and spatial databases for forest management and preparation of working plan. GIS technologies have been providing foresters with powerful tools for record keeping, analysis, and decision making. The use of GIS in forest management is significant through which forestry sectors can maximize their benefits.

1.4.9 Water Resources

Water resources are at the heart of sustainable development in many regions of the world. Water of sufficient quantity and quality is an essential resource for agriculture, industry, and tourism, but also for everyday life in cities and villages. A growing population and a growing economy not only lead to increasing demand for water, they also cause increasing pollution that may threaten the very water resources. This growth depends on and thus threatens the sustainability of development. Because water in its occurrence varies spatially and temporally, its study using GIS is especially practical. GIS platforms are becoming increasingly dynamic, narrowing the gap between historical data and current hydrologic reality. GIS has many application fields such as flood mapping and monitoring, hydrological modelling, rain water harvesting, management of

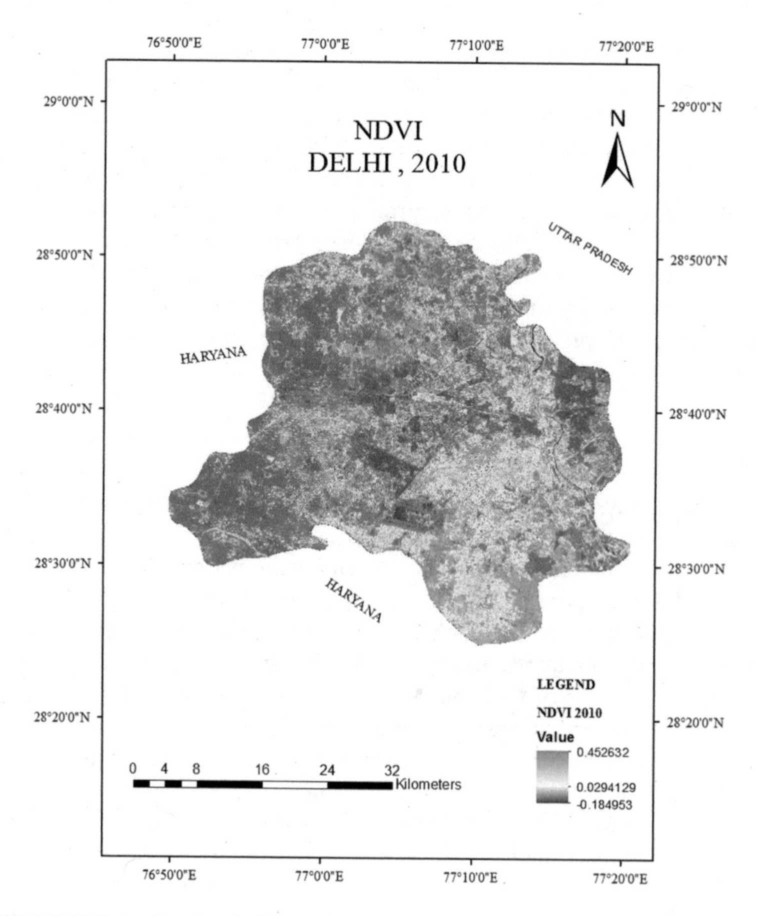

FIGURE 1.7 GIS Application in Forest Studies.

irrigation water system, ground water exploration, monitoring of drought, monitoring water pollution activities, Shallow Slope Stability (SHALSTAB), stratigraphy, 3D borehole, drastic, hydrostratigraphy, groundwater plume, aquifer recharge, mudflow, darcy flow, etc. (Fig. 1.8). Overall, four new and rapidly evolving sets of tools with strong connections to water resource applications of GIS can be identified. These are spatial interpolation tools, watershed delineation, flow tracing tools, map algebra tools, computer cartography and visualization tools.

1.4.10 Coastal Development and Management

The coastal zone represents varied and highly productive ecosystem such as mangrove, coral reefs, see grasses, and sand dunes. The coasts all around the world

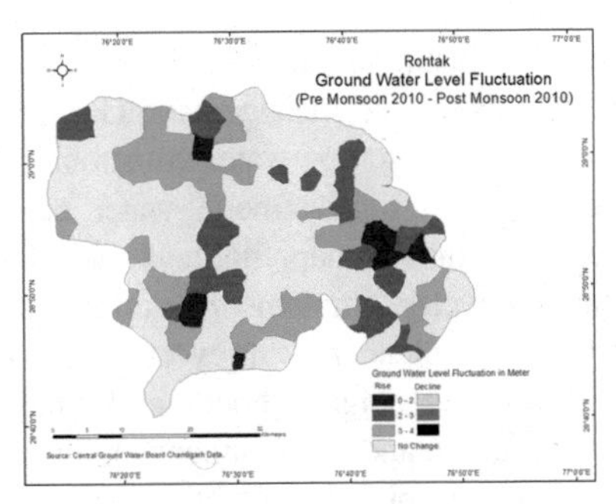

FIGURE 1.8 GIS Application in Water Resource Studies.

are fast developing and firm management policies have to be established. For any management of the shore to be effective, it is necessary for the policies to be based on informed decision making. As with any other GIS application, the data involved in creating a coastal GIS database fall into a number of distinct categories (i.e., basic geodetic/planimetric data, topographic data, qualitative and quantitative attribute data, time series data, and metadata). GIS can be applied in order to keep track of a wide range of natural and human-induced changes, including changes in the extent and ecology of coastal areas, analysis of erosion and shoreline changes, assessment of potential and actual flood hazard and damage, ocean water intrusion, silting up of harbors, and the effectiveness and impacts of mitigation efforts such as dredging, monitoring the changes of land use in the coastal hinterlands, in particular the growing urbanisation of the coastal fringe, and monitoring the behavior of oil spillages in coastal environments.

1.4.11 Other Applications

As the utility sector is expanding at a great pace, the importance of GIS mapping has been realized the world over. In the **power sector**, GIS is playing its role in transmission (assets, plan routes, maintain rights-of-way, track vegetation management), distribution (outage management, customer service, field operations, situational awareness, accounting, planning, network operations) and generation (environmental impact assessment, suitable sites for power generation companies, preparation of baseline data for the project area such as indexing of the consumers), distribution, and important role in preparation of baseline data for the project area such as indexing of the consumers, metering of distribution transformer, feeders, mapping of all the electricity assets, and distribution of network over the entire assigned landscape. Location is fundamental to **telecommunication** services in mobile and wired telecom networks. Advantages of GIS in the telecom sector are to generate high-resolution 3D building models for accurate planning, mapping, and real-time spatial analysis for new market opportunities, and reveal ways to maximize operational performance. In the **health sector**, GIS provides a cost-effective tool for evaluating interventions and policies potentially affecting health outcomes. In the GIS domain, health data is helpful in explaining disease patterns of relationships with social, institutional, technological, and natural environments. In the sector of **waste water** disposal and management, GIS can help in generating layers for demand modeling, forecasting, source water vulnerability mapping and analysis, water-loss tracking, analysis and management, water conservation management, planning for water reuse, consumption mapping, and analysis. Every day, planners use GIS technology to research, develop, implement, and monitor the progress of their plans. GIS provides administrators, planners, surveyors, and engineers with the tools they need to design and map their neighborhoods and cities. Planners have the technical expertise, administrative savvy, and fiscal understanding to transform a vision of tomorrow into a strategic action plan for today, and they use GIS to facilitate the decision-making process.

Concluding Remarks

Development of spatial analysis tools and technologies makes an easy way to describe, analyze, and predict various problems faced by mankind today. GIS is an important tool of geospatial analysis, which is now widely used in a range of sectors such as in spatial crime mapping, weather forecasting, land suitability analysis, traffic management, mapping health epidemics, crop monitoring, telecommunication, etc. GIS as a tool of spatial analysis continuously evolved from merely just a mapping tool to have unique modeling and analytic techniques such as artificial intelligence, network analysis, surface terrain modeling, digital elevation modeling and autocorrelation, etc. The ability to collect and analyze data in real time brought GIS to a whole new level. With the help of freely available open-source GIS softwares, access to the latest tools and techniques of spatial analysis for public welfare would increase its utility.

REFERENCES

Andresen, M. A. (2009) *The Place of Environmental Criminology within Criminological Thought*. Taylor and Francis, Oxfordshire, U.K. https://www.researchgate.net/profile/Martin_Andresen2/publication/272944229.

Bao, S. (1999) An overview of spatial statistics. In Alessandra, P., Nicola, S., and Chiara, S. (2003) *The Application of a Spatial Regression Model to the Analysis and Mapping of Poverty, Environment and Natural Resources Service No. 7, Sustainable Development Department*. University of Michigan, USA, China Data Center. http://www.fao.org/3/y4841e/y4841e00.htm#Contents.

Brantingham, P. J. and Brantingham, P. L. (1991) *Environmental Criminology*. Waveland Press, Prospect Heights, IL. https://en.wikipedia.org/wiki/Environmental_criminology.

Borneman, E. (2014) *Recent Advances in GIS Technology*. GIS Lounge. https://www.gislounge.com/recent-advances-gis-technology. In UK Diss.com. https://ukdiss.com/examples/geographic-information-system.php.

Camara, G., Monteiro, A. M., Fucks, S. D., and Carvalho, M. S. (2004) *Spatial Analysis and GIS: A Primer* (Online). https://www.researchgate.net/publication/2934461_Spatial_Analysis_and_GIS_A_Primer.

Clark, P. J. and Evans, F. C. (1954) Distances to nearest neighbor as a measure of spatial relationships in populations. *Ecology*, 35(4), 445–453. In Paramasivam, C. R. and Venkatramanan, S. (2019) *GIS and Geostatistical Techniques for Groundwater Science*. https://www.sciencedirect.com/topics/earth-and-planetary-sciences/spatial-analysis.

Cohen, L. and Felson, M. (1979) Social change and crime rate trends: A routine activity approach. *Am. Sociologic. Rev.*, 44, 588–608. http://faculty.washington.edu/matsueda/courses/587/readings/Cohen%20and%20Felson%201979%20Routine%20Activities.pdf.

Cucala, L., Genin, M., Occelli, F., and Soula, J. (2018) A multivariate nonparametric scan statistic for spatial data. *Spatial Stat.*, 29, 1–14. In Paramasivam, C. R. and Venkatramanan, S. (2019) *GIS and Geostatistical Techniques for Groundwater Science*. https://www.sciencedirect.com/topics/earth-and-planetary-sciences/spatial-analysis.

Fotheringham, S., Brunsdon, C., and Charlton, M. (2002) Geographically weighted regression: The analysis of spatially varying relationships. Wiley, Hoboken. In Blachowski, J. (2016) *Application of GIS Spatial Regression Methods in Assessment of Land Subsidence in Complicated Mining Conditions: Case Study of the Walbrzych Coal Mine (SW Poland). Nat Hazards*, 84, 997–1014. https://doi.org/10.1007/s11069-016-2470-2.

Getis, A., Getis, O., and Keith, J. (1992) The analysis of spatial association by the use of distance statistics. *Geog. Anal.*, 24, 189–206. In Alessandra, P., Nicola, S. and Chiara, S. (2003) *The Application of a Spatial Regression Model to the Analysis and Mapping of Poverty, Environment and Natural Resources Service No. 7 Sustainable Development Department*. http://www.fao.org/3/y4841e/y4841e00.htm#Contents.

Georges, D. E. (1978) The *Geography of Crime and Violence, Resource Papers for College Geography no. 78.1*. Association of American Geographers, Washington D.C. https://shodhganga.inflibnet.ac.in/bitstream/10603/205379/6/06_chapter_1.pdf.

Griffith, D. A. (2011) *Spatial Autocorrelation and Spatial Filtering: Gaining Understanding through Theory and Scientific Visualization*. Springer, Berlin. GIS Lounge. https://www.gislounge.com/gis-spatial-autocorrelation/.

Guerry, A. M. (1833) Essai sur la Statistique Morale de la France. France, Crochard. In Ksenija, B. and Jelena, M. (2017) *Geography of Crime and Geographic Information Systems*. J. Foren. Sci. Crim. Investig., 2(4), 55559. https://juniperpublishers.com/jfsci/pdf/JFSCI.MS.ID.555591.pdf.

Jeffery, C. R. (1971) Crime prevention through environmental design. In Martin, A. A. (2009) *The Place of Environmental Criminology within Criminological Thought*. Sage Publications, Beverly Hills, CA. file:///C:/Users/dell%20vostro/Downloads/Andresen_Introduction_Place_EC_Criminology % 20 (1).pdf.

Ksenija, B. and Jelena, M. (2017) Geography of crime and geographic information systems. *J. Forens. Sci. Crim. Investig.*, 2(4), 55559 0021. https://juniperpublishers.com/jfsci/pdf/JFSCI.MS.ID.555591.pdf.

Newman, O. (1976) Design Guidelines for Creating Defensible Space. U.S. Government Printing Office, Washington, DC. In Andresen, M. A. (2009) *The Place of Environmental Criminology within Criminological Thought*. https://www.researchgate.net/publication/285696892_The_place_of_environmental_criminology_within_criminological_thought.

Oliver, M. A. and Webster, R. (2007) Kriging: A method of interpolation for geographical information systems. *Inter. J. Geograp. Inform. Sys.*, 4(3), 313–332. In Paramasivam, C. R. and Venkatramanan, S. (2019) *GIS and Geostatistical Techniques for Groundwater Science*. https://www.sciencedirect.com/topics/earth-and-planetary-sciences/spatial-analysis.

O'Sullivan, D. and Unwin, D. (2010) *Geographic Information Analysis* (2nd ed). John Wiley & Sons, Hoboken, NJ, p. 167. GIS Lounge. https://www.gislounge.com/gis-spatial-autocorrelation/.

Quetelet, M. A. (1831) Research on the propensity form crime at different ages. Cincinnati Anderson, USA. In Ksenija B. and Jelena M. (2017) Geography of crime and geographic information systems. J. Forens. Sci. Crim. Investig., 2(4), 55559. https://juniperpublishers.com/jfsci/pdf/JFSCI.MS.ID.555591.pdf.

Shaw, C. R. and McKay, H. D. (1942) *Juvenile Delinquency in Urban Areas*. University of Chicago Press, Chicago. In Ksenija B. and Jelena, M. (2017) Geography of crime and geographic information systems. J. Forens. Sci. Crim. Investig., 2(4), 55559. https://juniperpublishers.com/jfsci/pdf/JFSCI.MS.ID.555591.pdf.

Treadwell, J. (2006) *Criminology*. SAGE. https://www.ukessays.com/essays/criminology/the-contribution-of-the-chicago-school-of-criminology-criminology-essay.php.

2 Spatial Crime Mapping

The evolution of crime mapping has heralded a new era in spatial criminology, and a re-emergence of the importance of place as one of the cornerstones essential to an understanding of crime and criminality. Early work on crime and place focused on offenders and where they lived and socialized; later work from the 1970s onwards focused on offenses and victims as in the Chicago School of thought. The causes and circumstances leading to emergence of crime and criminals in the society are very complex. These may include family violence, neglect, unhealthy conditions at school, poor family background, lack of social opportunities, violent neighborhood, and harsh environmental conditions. The connections between poverty, place, and crime are still much debated by criminologists. Today, urban spaces are marked by the contours of financial opportunities and class-based social processes. Crime in most urban areas is mostly determined by the types of housing, patterns of vulnerable public places, street police patrolling, and the neighborhood. To prevent the re-occurrences of crimes it is essential to understand the levels of crime in the society. The occurrence of crime shows strong spatial variations with mapping and explaining patterns of crime. The major emphasis in criminology is on the analysis of cause and impact of crime in the society and to identify the conditions which leads to crime in the society. Three theories are predominant while studying the spatial criminology as: crime pattern theory, rational choice theory and routine activity theory. An important geographical measure, the location quotient provides significant measurement of crime rate. It is a ratio of the percentage of a crime type in a subregion relative to the percentage of that crime type in the region as a whole. Newman (1972) identified the relationship between specific aspects of urban design and levels of crime. Newman's own reworking of defensible space theory (1996) stresses the need to move beyond urban design to address community relations. Another tool of crime mapping is geographic profiling, which prioritizes lists of potential offenders and helps in corresponding investigation. A key factor in the research of the influence of the environment on the level and nature of crime is the assumed distance of influence or micro-place. It is clear that criminal events occur when a motivated offender and a suitable target converge in time and space. So, it is essential to have a step-wise statistical analysis and visualization process to identify general areas affected by high criminal activity popularly known as crime hot spot mapping and to focus on these areas. Another method is spatial-temporal scan statistic, which is used to detect crime clusters in space and time.

2.1 INTERACTIONS BETWEEN CRIME, SPACE, AND SOCIETY

Scientific knowledge regarding spatial patterns of crime is very essential nowadays for the containment and improvement of safety at public places in urban areas. Analysis of spatial criminology has become key area of research in various disciplines such as in geography, sociology and environmental criminology. In recent decades, the crime level has touched the alarming levels. So, it becomes essential to have a careful and in-depth study of causes and trends of crime at different levels. The methods and policies to control the increasing crime needs more attention nowadays. Effective measurement and comparative developments in crime and justice trending from a global perspective remains a key challenge for international policy makers (Mark et al., 2003). The evolution of crime mapping has heralded a new era in spatial criminology, and a re-emergence of the importance of place as one of the cornerstones essential to an understanding of crime and criminality. While early criminological inquiry in France and Britain had a spatial component, much of mainstream criminology for the last century has labored to explain criminality from a dispositional perspective, trying to explain why a particular offender or group has a propensity to commit crime (Jerry, 2014).

Four major stages have been identified by social scientists in which the spatial and ecological perspectives on crime underwent during the past 150 years. First, there was the Cartographic School of Criminology, which started in France, and spread to other European countries, most notably England, from 1830 to 1880. The second stage was the Chicago Ecological School of the 1920s and 1930s, followed by a third stage, the factor analytic school of the 1950s. The final and current stage is the geography of crime and environmental criminology (Lawman, 1982). Crime and the fear of crime are significant aspects of daily life and as such have been studied closely by human geographers who have examined the interactions between crime, space, and society. The occurrence of crime shows strong spatial variations and, perhaps unsurprisingly, work by geographers was initially concerned with mapping and explaining patterns of crime. The 1980s and early 1990s were a rich period for the geography of crime, with several important books emerging that included Evans and David Herbert's (1989) *The Geography of Crime*, Herbert's (1982) *The Geography of Urban Crime;* and Evans, Fyfe, and Herbert's (1992) *Crime, Policing, and Place* that set the agenda for research into the geographies of crime in this period (Fyfe, 2000).

Social life is conducted in social space. It is very essential to consider the range of different spaces where we spend time in or pass through on a typical day: home, street, college, workplace, shop, library, bar, sports center, cinema, friends' home. Consider by what means and at what time of the day we journey between the spaces and by the means as on bike, on foot, car, bus, and train. Each space has its own internal rules of conduct, breaching these rules can create potential for deviance. Geographers argue that the spaces are not simply the backdrop for our social interactions. but by contrast they help to shape the very nature of our social interactions. In other words, space has the power to shape social life.

What does it mean to think spatially about crime? A first stage is to ask where recorded crime takes place in addition to asking who commits crime and why.

This helps to build up a profile of the places or environments where most crime and control encounters occur, along with profiles of offenders and victims. Criminologists began to focus on these issues in earnest in the 1970s, taking the advantages of the earlier work of the Chicago School. Criminal justice practitioners now make routine use of crime mapping to allow them to observe spatial patterns.

A second stage is to consider how places can be altered in ways that might reduce crime. This can involve a number of actors, from definitions of what makes a particular location crime-prone or safe to the arrangement and purpose of buildings to local beliefs or memories about a place. It can also involve a number of agents, from planners, developers, and politicians, who have the power to change spaces, to ordinary people who have the everyday task of negotiating existing spaces. A third stage is to consider how we come to know about space and crime in the first place and what we do with that knowledge. Mapping the spatial criminology is very essential and an important tool in the field of criminological research (Fig. 2.1). This necessarily raises questions about the source of the statistics and the nature of mapping technology. More recently, the importance of the Internet in both areas has raised the issue of global public access to this kind of geo-data.

2.1.1 Offenders, Offenses, and Places

Offenders are the people who disobey the justice delivery system and indulge in activities that are harmful to the society. The causes and circumstances leading to emergence of crime and criminals in the society are very complex. These may include family violence, neglect, unhealthy conditions at school, poor family

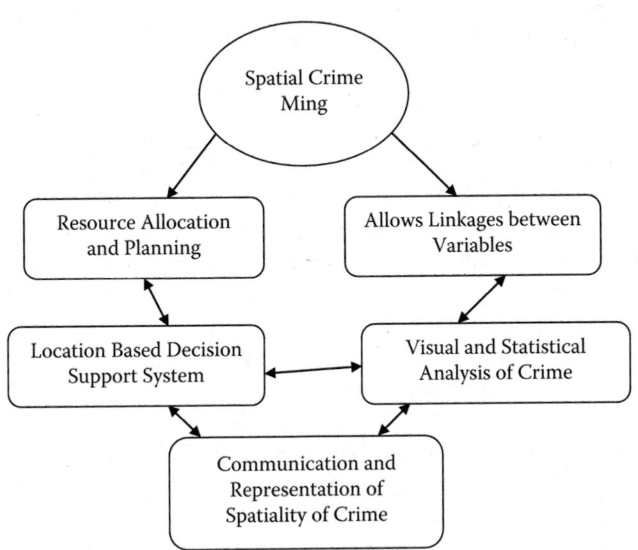

FIGURE 2.1 Spatial Crime Mapping.

background, lack of social opportunities, violent neighborhood, and harsh environmental conditions. Chicago School sociologists used terms such as "urban sociology," "human ecology," and "ecology of crime." In the 1970s and 1980s, "environmental criminology" was used until similar terms began to be more frequently used in relation to green issues. Around the same time, many criminological debates moved away from a traditional focus on the causes of crime to a post-welfare focus on crime prevention and management. As a result, emphasis has been started on situational crime prevention (SCP) and crime prevention through environmental design (CPED). Mike Davis explores crime and control in Los Angeles as an extreme example of the ecology off ear (1999). Others refer to socio-spatial criminology (Bottoms, 2007) or crime and community. On the more quantitative cartographic side, geo-criminology and crime mapping (Vann and Garson, 2001) are more frequently used.

Park and Burgess's work in Chicago in the early 20th century stressed the relationship between urban environment, actions, and values. They saw social science as a form of human ecology (1925). Burgess's zonal theory of urban development suggested that Chicago and other large cities were structured around five concentric circles. The non-residential central business district was surrounded by the zone in transition, an area of cheap rented housing attracting different generations of migrants. Next came three residential areas of increasing affluence. Other Chicago scholars built on this model. Shaw and McKay's (1942) studies of juvenile delinquency showed that a very high proportion of young offenders had grown up in the zone in transition. They explained this as an effect of the social disorganization, which characterized this area. A churning migrant population with shifting moral values, high levels of poverty, and low levels of community cohesion produced teenagers prone to commit crime. More recent U.S. criminological research linking a community's crime levels to its capacity for collective efficacy has some clear links to these early Chicago studies.

If early work on crime and place focused on offenders and where they lived and socialized, later work from the 1970s onwards focused on offenses and victims (Bottoms, 2007). One study argued that offenders tended to commit crime in an area which is culturally familiar to him in some way but not generally his own neighborhood. Victim surveys allowed area victimization rates (the level of offenses against a particular group in a particular area) to be compared with area offenserates (all recorded offenses in a particular area). Other studies, such as Baldwin and Bottoms (1976) on Sheffield, questioned the link between offenders and zones in transition in the UK context, stressing instead the importance of the housing market in shaping community relations.

The connections between poverty, place, and crime are still much debated by criminologists. Bottoms (2007) reviews recent studies exploring the link between deprivation and offender rates. Communities with high levels of collective efficacy, or high levels of cohesion and mutual trust, will be willing to intervene to challenge behavior in a given setting and stop it from escalating. Communities with low levels of collective efficacy, rather like those Burgess defined as living in the zone in transition, will be less willing or able to intervene.

2.1.2 Crime Prevention, Space, and Communities

Crime prevention may be defined as the process which addresses crime problems at their source of origin rather than waiting for the judgment of criminal justice system and punishing the criminals. Developing crime-fear-free public space should be the core objective of planners and policy makers from local to global levels. Today, urban spaces are marked by the contours of financial opportunities and class-based social processes. Crime in most of urban areas is mostly determined by the types of housing, patterns of vulnerable public places, street police patrolling, and the neighborhood. Crime and place, as the discussion so far has shown, are now very firmly discussed in relation to community and area dynamics. This raises the further question—is it possible to change these dynamics and the places in which they are played out in order to reduce crime? Put another way, can crime be cut by reshaping a space or altering community relations in a space or both? Criminological work on this first focused on changing spaces but has since moved on to focus on changing community relations.

2.1.3 Understanding Levels of Crime

To prevent the reoccurrences of crimes, it is essential to understand the levels of crime in the society. To understand, the levels of crime emphasis should be given on the authentic source of data. Generally three types of standard source of data are used to understand the levels of crime in the society. These are crime reports obtained from concerned offices, self-reports, and victim survey reports. The nature of reporting of crime varies from country to country, depending on the type of crime. In India, the National Crime Records Bureau collects crime data from the police headquarters of all the states across the country and has a system to standardize the data it receives. It categorizes the reports mainly into offenses against the person and offenses against property (Chockalingham, 2003). So, the approaches to deal with social crime should have a major focus on the causes of crime at the base level, which may have their origin in unhealthy social environment, unemployment, social isolation, etc. United Nation Habitat's Safer Cities Program 1996 also stressed certain pillars of crime prevention in urban areas. This program emphasized on integrated approach including multilevel governmental and sectoral involvement and holistic activities to deal with the problem (UN-HABITAT, 2007).

2.1.3.1 Community Level

- Develop integrated youth policies.
- Identify people at risk.
- Develop crime-free childhood.
- Public awareness through various channels.
- Organizing seminars, talks, and group discussions regarding crime at public place.

2.1.3.2 International Initiatives

- More need to design and develop international programs such as Global Network on Safer Cities (GNSC) of UN-HABITAT.
- UN Network of Institutes to address crime prevention and criminal justice at the global, regional, and subregional levels.
- The Education for Justice (E4J) initiative seeks to prevent crime and promote a culture of lawfulness through education activities.
- Developing crime solving approaches/models as per Doha Declaration such as SARA and Ekbloms 5Is.

2.1.3.3 Designing and Housing

- Designing the public space in a way to reduce the opportunity to commit crime.
- Proper street layout and lighting.
- Proper electronic surveillance at public place.
- Community participation in planning and developing public places.

2.1.3.4 Law Enforcement

- Good behavior of law enforcement agencies with the victim of crime.
- Problem-oriented law enforcement policies.
- Organize events to boost public trust in policing.
- Organize police/public meet to identify the core areas.
- Strict adherence to law enforcement policies.
- Quick response and counseling of affected and the offenders also.

2.1.4 Crime and Surveillance

Surveillance is an important feature of situational crime-prevention measures that seek to reduce opportunities for offending through "eyes on the street," target hardening, and environmental management. Authors such as Oscar Newman and Alice Coleman advocated designing the urban environment to improve natural surveillance. Coleman's work was controversial as it implied a link between building design and crime, rather than the social conditions that underpinned a neighborhood. Other authors, such as Jane Jacobs, advocated closer community co-operation to facilitate what she referred to as natural webs of surveillance formed of people living and working in neighborhoods (Richard, 2015).

2.1.5 Changing Spaces: Urban Design and Crime

The spatial distribution of crime incidents varies in accordance with type. The most obvious difference is between urban and rural areas with a much wider range of crimes occurring in urban environments (Esteves, 1995; Ferreira, 1998). It was

Newman (1972) who first identified the relationship between specific aspects of urban design and levels of crime. In his theory, Crime Prevention through Environmental Design (CPTED), he argues that urban design influences the incidence of crime and the formation of hot spots. U.S. architect Oscar Newman (1972) used the concept of "defensible space" in the 1970s to argue that it was possible to modify the built environment to reduce the opportunity for crime and to promote community responsibility. Newman's idea, which centered on public housing design, helped to shape new approaches within what was then still referred to as environmental criminology. Situation crime prevention (SCP) and crime prevention through environmental design (CPED) advocated changes in physical environments and physical objects within them. These strategies have gradually become part of everyday life in public, residential, commercial, and financial urban sectors. Street fixtures such as benches, bus shelters, playgrounds, and lighting were all increasingly designed to screen out undesirable activity. Surveillance equipment and CCTV were used not only to monitor but also to deter wrong-doing.

Urban design and surveillance were taken up by academics and planners at a time when the old focus on the causes of crime was beginning to give way to a new focus on the need to manage crime. SCP and CPED, for example, are clearly linked to Felson's "routine activity theory." Crime has been seen as an inevitable phenomenon that can best be managed by reducing the opportunity to commit an offense rather than by seeking to reduce individuals' desire to commit a crime in the first place.

Criminology remains divided on the implications of this shift. Some argue that it addresses the needs of, and empowers, those communities—often among the most deprived—that live with the realities of high crime rates. Community safety is identified as an important element in any kind of neighborhood regeneration. Mike Davis (1999) offers an extreme but very interesting view here. His account of Los Angeles as an "ecology off ear" reworks the original Chicago School zonal theory and argues that the linking of urban design and policing has led to a destructive militarization of urban landscapes which protects privilege and punishes poverty.

Newman's own reworking of defensible space theory (1996) stresses the need to move beyond urban design to address community relations. People should feel that they own public space and share a responsibility for it not simply that they are being monitored. This kind of thinking is evident more in the communitarian approaches to governance that emerged in the 1990s and which are also very much linked to post-welfarism. New Labor's 1998 Crime and Disorder Act has had a major impact on British approaches to crime and community. Crime was to be tackled not just by the police and the courts but by new Crime and Disorder Reduction Partnerships (CDRPs) which were set up in over 300 local authorities. The partnerships require a multi-agency approach, typically involving the police, local councils, health authorities, and voluntary agencies. The emphasis is on identifying both local crime problems and what works to reduce these. There are two key strategies here. First, the community is made responsible as part of a wider dispersal of power and, second, these new styles of local policing encourage a new kind of attention to local trouble spots.

Increasing interest in the localized nature of crime has led to highly localized policing strategies and even localized criminal justice legislation. Dispersal orders, ASBOs, curfews, and other measures are all tailored to particular environments.

They aim to stop certain people behaving in certain ways in certain spaces at certain times. Civil rights campaigners have warned that these spatial techniques represent a dangerous trend because, among other things, they sanction a move away from the principle of a common, universal criminal justice system operating equally across a state. They have launched a number of legal challenges to the government on these issues.

Some criminologists believe that this kind of work is valuable because it can show more precisely where crime problems are and where polices should target their resources. Echoing the earlier discussion of the night time economy, Bromley and Nelson's study of alcohol consumption and crime in a British city concludes that a detailed knowledge of the variety of spaces and times of alcohol-related crime and disorder is key to the development of appropriate urban design, planning, and licensing policies and can be used to inform a more closely targeted policing strategy. White and Sutton (1995) stress the limits of quick fixes for crime, arguing against episodic initiatives and technological strategies in favor of strategies "which see crime and public safety as stemming first and foremost in the community." Herbert and Brown (2006) argue, similarly, that the relationships between identities, values, and spatial environments are complex and should not be oversimplified (Carrabine et al., 2009). Finally, this kind of analysis may work for public order offenses, but does not help the police to tackle other kinds of crime which take place in private, as opposed to public, space: white-collar crime, domestic violence, fraud and state crime, for example.

2.2 SPATIAL PROCESSES AND CRIMINOLOGY

Criminology may be defined as the integrated study of criminals including their behavior and activities and the response of the society facing these crimes. So, the major emphasis in criminology is on the analysis of cause and impact of crime in the society and to identify the conditions which leads to crime in the society. Crime and its spatial process have been studied for close to 200 years since the work of Andre Michel Guerry (1833) and Adolphe Quételet (1842), French and Belgian, respectively. Their works were not only important for spatial criminology, but positivist criminology, more generally. Since this time, a large body of research has emerged which investigates the spatial dynamics of crime at a variety of scales and contexts.

Criminologists conceive and measure spatial influence in a number of different ways. According to some criminologists differences in rate and pattern of crime is due to the aerial characteristics or more geographically aerial differentiation. Thus, the spatial structure of areas and their respective adjacency to other areas become the central points of interest. Another body of spatial research in criminology focuses on individuals as the units of analysis, who themselves are located within spatially oriented structures such as gangs, schools, or neighborhoods (Townsley, 2009).

There are many dimensions of spatial criminology that are of particular interest at this time, including understanding the role of police foot patrol and car patrol in reducing crime, understanding the crime drop of the 1990s, furthering the understanding of relationships such as unemployment and crime, and helping to understand the transition economies in Central and Eastern Europe. In order to understand and interpret the spatial dynamics of crime in different regions, it is essential to emphasis on four key issues: spatially referenced crime rates, the location quotient, geographic profiling, and crime and place. With the help of these issues it would be easy to visualize and analyze the movement patterns of criminals and victims.

2.2.1 Spatially Referenced Crime Rates

In order to calculate a crime rate, there is a requirement of two types of data: crime data and population at risk data. In a defined geographical area, the number of criminal events in a specific time period may be defined as the crime count. This count is mostly based on different data sources such as official records of crimes and crime survey reports, etc. The most well-known official crime data are the Uniform Crime Reporting (UCR) data. These data had its beginning in the United States in 1930 and now nearly all law enforcement agencies in the United States provide data to the Federal Bureau of Investigation for the UCR counts (Mosher et al., 2011). In Canada, the UCR began later, in 1962, and is gathered by the Canadian Center for Justice Statistics. Responding to the UCR survey in Canada is mandatory. In 1988, a new version of the UCR in Canada was started that added information regarding incidents, victims, and accused persons. Equivalent data in the United States has been collected under the National Incident-Based Reporting System (NIBRS), that began in 1987 (Federal Bureau of Investigation, 2012). In India, the National Crime Record Bureau is the nodal agency for collecting and analyzing of the data regarding various crimes types and issues in the country. Another form of official crime data is calls for service data from the police. Though often considered unofficial because these data are not dependent upon a criminal charge and often represent police activity, these are very useful data because a geographic location and time is most often included and the calls for service can be separated to only include the calls that the police identify as a criminal event (Sherman et al., 1989).

Though the official data regarding crime in a region is sufficiently available but still there are certain limitations. There are possibilities of variance in the reporting of crime from various agencies because different agencies have their own pattern of data collection and analysis. Crime reporting can vary from police department to police department. This may be due to factors such as the local policing and/or population culture considering factors such as a lack of tolerance on particular issues. Second, no major criminal events has been reported to the police. In Canada, only 31% of criminal victimization was reported to the police in 2009, down from 34% in 2004 and 37% in 1999 (Perreault and Brennan, 2010). There would be difficulties in analysis and implementation if the spatial patterns of crime for official data would be different from the spatial patterns of crime. The desired output such as mapping the crime hot spots would be spurious.

Data regarding population at risk would be essential to calculate the crime rates. In spatial criminology, that often uses the census tract as a spatial unit of analysis, the most common source of population at risk data is the census resident population. The data is very easy to obtain and have high quality, but there are two issues while using the data. First, the exercise of census is only undertaken every 5 (Canada) or 10 years (United States, United Kingdom, and India) in most countries. Consequently, depending on the date of the crime data obtained there may be as many as a 9-year discrepancy between crime count and population at risk data. If population is stable over long periods of time then it wouldn't be problematic, per se, but there is considerable growth in cities which may lead to significant overestimates of crime rates if crime counts are for a later year than the resident population.

The second issue while using the resident population from a census is that population at risk is whether or not an accurate representation of the total population. This is a critical question in spatial criminology because by using the crime rate to represent an overall measure of risk the variable representing the population at risk should measure where people actually are located (Boggs, 1965). Recent research on this issue has shown that the ambient population can be quite different from the resident population leading to changes in inference regarding spatially referenced crime rates (Andresen, 2006). Data from the Oak Ridge National Laboratory was used to measure the ambient population that was defined as the number of people in a square kilometer averaged over a 24-hour period for any typical day of the year.

Ratcliffe and Taniguchi in their investigation in Camden, NJ from 2005–2006 searched for drug gang corners and their influence on violent crime and property. They used Intensity Value Analysis (IVA) and Thiessen polygon analysis and found that the crime level around drug gang corners is higher, but their influence is spatially differentiated. In an investigation of the similarity between maps of resident-based and ambient-based crime rates, Andresen and Jenion (2010) found that, despite the visual appearance of similar spatial patterns and a statistically significant relationship, resident-based crime rates were very poor predictor of ambient-based crime rates. Though we do not discuss this prediction, this is what we assume when we use the resident population. It may not be the best measure, but it is representative of the true population at risk. Moreover, places that attract populations throughout the day, measured using the ambient-resident population ratio, are commercial/shopping areas and major transportation routes.

In subsequent research, Andresen (2011, 2013) showed that ambient-based crime rates exhibited significant differences to resident-based crime rates in the context of local analysis and local Moran's. In a spatial regression context, the similarity of results for resident and ambient-based crime rates differed on the year of data and the size of the unit of analysis. In some cases, the results from the ambient-based crime rates outperformed the resident-based crime rate results in the context of greater (pseudo) R^2 values, more reasonable parameter estimates, more statistically significant relationships, and a better concordance with theoretical expectations. In other cases, there were very few differences with limited consequences to the results.

2.2.2 A Spatial Alternative Measure of Crime: Location Quotient

An important geographical measure, the location quotient, Branting provides significant measurement of crime rate. The calculation of crime rates has two sources of measurement issues: the criminal event counts and the population at risk. There is an alternative to the crime rate that only relies upon crime event data, however, that has been used in spatial criminology for 20 years but is arguably under-utilized, the location quotient. It is essential to measure the level of representation based on some spatial unit of analysis. For this the method of location quotient is one of the important tools to analyze the representation of an activity of a particular area in relation to the entire study area.

The location quotient is a ratio of the percentage of a crime type in a sub-region relative to the percentage of that crime type in the region as a whole. The value of location quotient indicates the relative representation of a particular phenomenon with reference to whole study area. If the location quotient is equal to one, the sub-region has a proportional share of a particular crime; if the location quotient is greater than one, the sub-region has a disproportionately larger share of a particular crime; and if the location quotient is less than one, the sub-region has a disproportionately smaller share of a particular crime. For example, if a sub-region has a location quotient of 1.20, that sub-region has 20% more of that crime than expected given the percentage of that crime in the region as a whole. Consequently, it can be said that this sub-region "specializes" in that particular crime type. In this way it becomes easy for the researcher to identify the patterns of crime in different regions.

First introduced to the criminological literature by Brantingham and Brantingham (1993) as a descriptive tool, the location quotient has been used by subsequent researchers to investigate a number of phenomena such as drug markets, theory testing, shootings, and automotive theft markets. It is an important alternative method to measure patterns of criminal activity. For example, as outlined in Andresen (2013), the spatial patterns of crime rates versus location quotients are somewhat similar for a number of crime types, but not able differences emerge when considering specialization versus risk. In the case of assault in Vancouver, Canada, a crime rate map indicates that this crime type is highly concentrated in Skid Row. However, the east side of the city (and the north east side in particular) exhibits a lot of assault specialization. In order to answer this question, both the crime rate and the location quotient need to be considered. First, the risk of an assault is greatest in Vancouver's Skid Row. But, given that an individual is going to be a victim of crime, it is more likely to be an assault on the east side of the city relative to the west side of the city. Similar differences in the spatial patterns of crime emerge for other crime types as well, but not always in the same places: robbery, sexual assault, theft (more specialization of this crime type is present on the western more affluent side of the city), theft from vehicle, and theft of vehicle.

Andresen (2013) also presented an inferential application of the location quotient (spatial regression) that produced some interesting results. Most important, the workhorse theories used in spatial criminology (social disorganization theory and routine activity theory) performed well in the context of crime specialization. However, because the dependent variable measured specialization not risk, the interpretations of the estimated parameters are different. In some circumstances, these differences are subtle such that the estimated parameters have the same sign, but how one interprets the results changes; in other contexts, the sign is opposite from the expectation in a crime rate context (a negative relationship between burglary and the unemployment rate, for example) that requires a completely different, but theoretically justified, explanation.

The LQC continues to be successfully applied to research on the spatial distribution of crime. Carcach and Muscat (2002) used the location quotients to analyze changes in crime structures over time and across geographical areas, and to examine the role that socioeconomic characteristics play in shaping the crime profile of areas.

In all of the applications of the location quotient, the researcher is forced to ask different questions with regard to the spatial pattern of crime, sometimes subtle and sometimes not. Asking these different questions, it is argued, allows the researcher to obtain a better understanding of the phenomena of crime. As such, the use of alternative measures of crime, particularly in a spatial context, should not be used to replace crime rates, but to supplement crime rates for a deeper understanding of the spatial dynamics of crime.

2.2.3 Geographic Profiling

One of the more well-known applications of the theories within spatial/environmental criminology is geographic profiling. Developed by D. Kim Rossmo, geographic profiling may be thought of as understanding the spatial dynamics of serial crime. Geographic profiling works on the assumption that a crime usually has a spatial dimension, which can be analyzed and used in a criminal investigation (Canter and Alison, 2000).

Geographic profiling is an investigative methodology that uses the location of a connected series of criminal events to determine them probable area of the offender's residence. Technically geographic profiling is used to identify the anchor point from which a serial offender bases his or her movements, but this is most often the residence. Because geographic profiling uses series of criminal events, the most common uses of geographic profiling are serial murder, rape, arson, and robbery. However, geographic profiling may also be used for non-serial crimes that involve multiple locations. The major function of geographic profiling is to prioritize a list of potential offenders to aid in a corresponding investigation. Geographic profiling has proven and continues to be a useful tool for the practice of investigative policing as well as the understanding of serial offenders from an academic perspective. Some of the more interesting recent geographic profiling applications have included: cellular telephone switch tower sites in kidnaping cases, the stores in which bomb components have been purchased, and an historical analysis of the locations of anti-Nazi propaganda postcards noticed in the streets of Berlin, Germany in the early 1940s (Rossmo, 2014). Additionally, there have been new applications of geographic profiling that have proven to be very instructive, but for different reasons (Rossmo, 2012).

The method of geographic profiling represents a special case of offender profiling. It is essential that the process of geographic profiling should be based on scientific procedures. It must be based on empirical observations of real crimes and the statistical models derived from them, which can then be verified or falsified. Standard geographic profiling uses the related locations in a series of criminal events to help identify the offender's residence (Fig. 2.2). Geographic profiling focuses the search for the offender using a combination of spatial behavior,

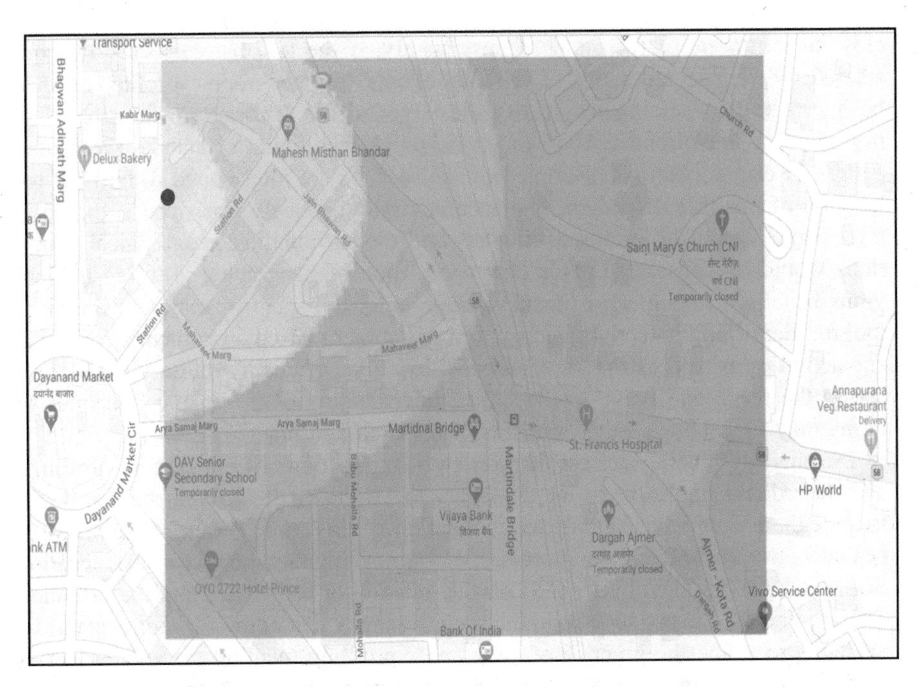

FIGURE 2.2 Geographic Profiling Technique to Locate Offender's Base.

environmental criminology theory, and algorithm. Using algorithm, a geo-profile of the offender, has been created which has been overlaid on the street map to identify the most probable area for the offenders' base. In the given map, the black dot represents the most probable point where the offender can be traced. Geographic profiling has also been applied to help military analysts to identify the location of enemy military bases (Brown et al., 2005), but it can also be used to identify the targets of terrorist attacks. The study of terrorist activities has identified the existence of "terrorist cell sites." These terrorist cell sites may be meeting places, rented apartments, etc., that are used as part of the planning of a terrorist attack. As it turns out, there is a strong spatial component to these terrorist cell sites. The research undertaken by Rossmo and Harries (2011) goes through how geographic profiling may be used to identify terrorist targets using these terrorist cell sites. Effectively, this use of geographic profiling is undertaken by taking the method of geographic profiling and turning it on its head. However, a number of factors may increase the level of accuracy. These factors may include spatial analysis of crime locations, distance from home to crime spot, gender and types of crimes, etc. Knowledge of these factors could enable an increase of accuracy of home location prediction (Manne, 2007). Rather than attempting to help identify the offender, geographic profiling is used to help identify the victim, whether that victim is a place or a person. As it turns out, terrorist cell sites are set up along the same principles as other offending activities. A non-terrorist offender tends to search for criminal targets in known areas, close to home but not so close as to reveal his or her identity. A set of terrorist cell sites tend to be organized close to but not too close to the target of the attack. Consequently, if the terrorist cell sites are considered the "connected series of criminal events" the geographic profile can then be interpreted as helping to identify the target of the terrorist attack rather than the home of the offender. It is an investigative technique to analyze the spatial pattern of crime locations or crime hot spots which are interrelated to find out the location of possible offenders. The accuracy may be easily improved if the basic information about the offense, offenders and the surrounding geographical conditions would be considered. This new application of geographic profiling has obvious benefits to counter insurgency and counter terrorism operations as another tool for identifying potential targets of terrorist attacks based on known terrorist cell site activities as well as the most probable locations of enemy military bases. This shows the reach and power of how the understanding of the spatial dynamics of crime may be useful in other domains of research and practice.

Despite the general interest of the newer applications of geographic profiling, and their direct implications for the safety and security of different human populations, there is another significant application of this more recent research. Most offenders usually undertake a short path of journey to target the victim. The location of the offense completely depends on the location of the victim and the offender. Though there are many serial criminal investigations that allow for more data to be available to refine the practice of geographic profiling, waiting for another set of serial homicides, for example, is not the most pleasant way of thinking of future data. If, and it does appear to be the case, that the predatory behaviors of other animal species are similar to ours, this new research can be used to better understand the predatory behavior of humans. For example, if humans, similar to sharks of the South African coast, also reduce the size of their search area as the age, this information may be useful for the ranking of potential offenders in a criminal investigation; in such a situation, it may be possible to solve a serial criminal investigation sooner because of a better understanding of spatial predatory behavior, more generally.

2.2.4 Crime and Place

As mentioned above, spatial criminology as we know it today began almost 200 years ago in France by Quételet and Guerry. One of the trajectories of spatial criminology over these many years has been the identification of spatial heterogeneity within the common spatial units of analysis of the day. For many years, the standard spatial unit of analysis within this literature has been the neighborhood or census tract. One of the great advantages of these units of analysis is the availability of census data for analysis such as with the crime rate and location quotient studies discussed above. Twenty-five years ago, the use of yet another smaller spatial unit of analysis began to emerge, called the micro-place, this literature is often referred to as the crime and place literature. These micro-places could be addresses, street intersections, or street segments, and widely used for citywide analyses because it became apparent that neighborhoods were far from being spatially homogeneous with regard to criminal activity.

A key factor in the research of the influence of the environment on the level and nature of crime is the assumed distance of influence. From a theoretical perspective in spatial criminology, the micro-place makes a lot of sense. Within routine activity theory (Cohen and Felson, 1979), a criminal event occurs when a motivated offender and a suitable target converge in time and space. That convergence, of course, does not occur in a neighborhood, but at a discrete location. In the geometric theory of crime (Brantingham and Brantingham, 1981), criminal events dominantly occur at our activity nodes and the pathways between them, but there are also particular places within these activity nodes where the majority of criminal events would occur. And lastly, rational choice theory (Clarke and Cornish, 1985) postulates that criminal events are the result of context specific choices. That context can vary significantly within a "neighborhood" such that the context of micro-places will be important for criminal decision making.

The first known research study to systematically investigate the micro-place in a city wide analysis was propounded by Sherman et al. (1989) who presented an interesting analysis of predatory crime in Minneapolis. In this study, Sherman et al. (1989) found that 50% of calls for police service were generated from 3% of street segments. This profound concentration of criminal events showed that even in the neighborhoods with the greatest criminal event levels, it had areas within its periphery that were crime free. Smith et al. (2000) investigated the integration of routine activity theory and social disorganization theory considering the micro-place rather than the more tradition census tract or other census defined unit of analysis. They found that the integration of these two theories was far more successful using the micro-place. Weisburd et al. (2004) and Groff et al. (2009) analyzed the micro-place in Seattle using trajectory analysis. Similar to Sherman et al. (1989), these authors found that 50% of all criminal events were accounted for by approximately 5% of street segments, over a 14-year period. Moreover, Groff et al. (2009) found that street segments with the same trajectory tended to cluster together.

2.2.5 Crime Hot Spots

One of the most common and innovative uses of crime mapping is to aggregate numerous crime events into hot spot maps. Modern tools of visual and spatial analysis, strongly supports the law enforcement and controlling agencies in their entire decision-making process concerned with crime. Crime records including crime history and crime pattern are valuable information required much for collective and integrated mapping of criminal events in spatio-temporal framework. Hot spot mapping of crime events is a common and innovative application of spatial crime mapping (Fig. 2.3). Areas on a crime map that are shaded with the same color are deemed to contain approximately the same density or frequency of crime. A step-wise statistical analysis and visualization process is essential to identify general areas affected by high criminal activity and to focus on these areas. Such as Lucy and Alexander (2014) used two packages of the R language: the Spatial Relative Risk and ggplot2 to analyze and visualize the concentration of crime, respectively, and estimated relative clustering of crime with Kernel Density Estimation (KDE).

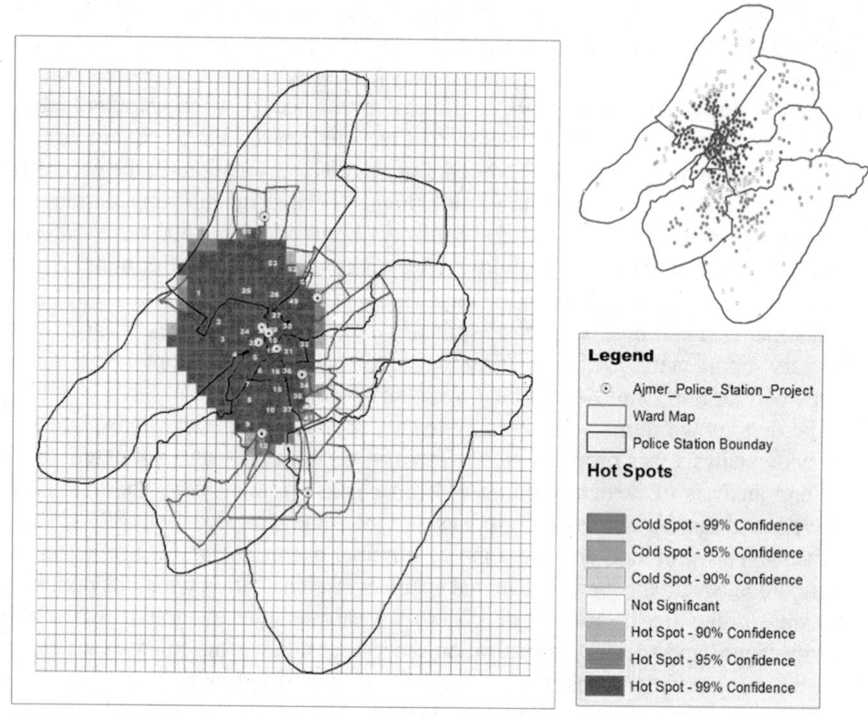

FIGURE 2.3 Crime Hot Spots.

Risk Terrain Modeling: Caplan and Kennedy (2016) developed a risk terrain modeling (RTM) as a statistically valid approach to identify spatially vulnerable places to crime through the creation of a risk of crime score based upon criminogenic features from the physical environment of an area. Caplan et al. (2011) indicated that RTM is more accurate at predicting crime than retrospective crime hot spot mapping. Anderson and Tarah (2018) applied the RTM to property crime victimization (i.e., residential burglary) in Vancouver, Canada.

Development of recent research in this area indicates the importance of the micro place in understanding the spatial dynamics of crime. Weisburd et al. (2013) in the context of crime prevention, expressed that over a 16-years' time period, 1% of street segments in Seattle represented 23% of all criminal activity across the city. This is an incredible concentration of criminal events in any context. In another set of analyses in Canada, Andresen and Malleson (2011) and Andresen and Linning (2012) revealed how concentrated crime can be in two Canadian cities, Vancouver and Ottawa. In Vancouver, in 2001, 50% of the criminal events under study were accounted for by just 5% of street segments an almost identical result compared to the work done on Seattle. This concentration varied by crime type from approximately 1% of street segments (sexual assault and robbery) to almost 8% (burglary). In Ottawa, in 2006, the concentrations were even greater: 50% of the criminal events under study were accounted for by 1.7% of street segments, ranging from

0.01% of street segments (commercial robbery) to 1.67% of street segments (burglary), the average is greater than the range because different crime types have their concentrations at different places.

In Vancouver, just over 60% of street segments have any criminal events whereas less than 10% of street segments in Ottawa have any criminal events. This is an incredible concentration within Ottawa that may be due to the limited number of crime types under analysis, but is noteworthy none the less. It is found that the percentage of street segments with any criminal events that account for 50% of criminal events as the base for the percentage, not all street segments in each city. These percentages represent concentrations within concentrations and indicate that there are clearly hot spots of crime within hot spots of crime and that spatial heterogeneity can even be present considering the micro-spatial unit of analysis.

2.3 SPATIAL CRIME THEORIES IN PRACTICE

There is no one cause of crime. Crime is a highly complex phenomenon that changes across cultures and across time. Activities that are legal in one country (e.g., alcohol consumption in the UK) are sometimes illegal in others (e.g., strict Muslim countries). As cultures change over time, behaviors that once were not criminalized may become criminalized (and then decriminalized again, e.g., alcohol prohibition in the USA). As a result, there is no simple answer to the question what is crime? and therefore no single answer to what causes crime? Different types of crime often have their own distinct causes. The theoretical and empirical work on research on crime may be traced back to the middle of the 19th century. The social ecology perspective evolved into more specifically focused place-based theories of crime, particularly the routine activities theory (Luc et al., 2000).

Since the 20th century, the interest in crime places continuously grows. The identification of crime hot spots was perhaps a watershed in refocusing attention on spatial/locational features of crime. This interest spans theory from the perspective of understanding the etiology of crime, and practice from the perspective of developing effective criminal justice interventions to reduce crime (Luc et al., 2000). Three approaches are predominant while studying the spatial criminology as: crime pattern theory, rational choice theory, and routine activity theory. According to the crime pattern theory, rationally and reasonably motivated offenders, during daily routine activities, are in contact with a relatively small part of the city. Among the perceived and unconscious nodes, paths and edges offenders select the appropriate objects or victims of a crime in a multistage decision-making process. The spatial distribution of crime in a city depends on its spatial pattern, land use, transportation system, and the street network (Natalia and Michael, 2017).

Here it becomes essential to provide an overview of some of the key criminological theories which seek to explain the causes of crime. Each of the theories covered has its own strengths and weaknesses, has gaps, and may only be applicable to certain types of crime and not others. The theories covered can be categorized as:

Biological theories
Sociological theories

2.3.1 Biological Theories

Biological explanations of crime assume that some people are "born criminals", who are physiologically distinct from non-criminals. The most famous proponent of this approach is Cesare Lombroso. Lombroso's most famous work, L'uomodelinquente (The Criminal Man), considered by many historians the founding text of modern criminology. Lombroso focused on the criminal rather than crime. Key concepts in the various editions of L'uomodelinquente were atavism, degeneration, and the idea of the born criminal (Beccalossi, 2010). Lombroso's work has long since fallen out of favor. However, biological theories have continued to develop. Rather than measuring physical features of the body, contemporary approaches focus on:

- Biochemical conditions (e.g., linked to poor diet or hormone imbalance)
- Neurophysiologic conditions (e.g., learning disabilities caused by brain damage)
- Genetic inheritance and/or abnormality
- Intelligence

These attempts, to locate the causes of crime within the individual, suggest that there are identifiable differences between offenders and non-offenders. In other words, the criminal is "other": in some way different or abnormal to everyone else.

2.3.2 Sociological Theories

Sociological approaches suggest that crime is shaped by factors external to the individual: their experiences within the neighborhood, the peer group, and the family. According to Shaw and McKay, the neighborhoods that have the highest rates of crime typically have at least three common problems: physical dilapidation, poverty, and heterogeneity. Shaw and McKay believed there was a breakdown of informal social controls in these areas and that children began to learn offending norms from their interactions with peers on the street. Thus, the breakdown in the conditions of the neighborhood leads to social disorganization, which in turn leads to delinquents learning criminal activities from older youth in the neighborhood.

Guerry and Quételet extensively interpreted and analyzed the spatial variation in community crime levels in terms of the varying social conditions of the local populations. The Chicago School of the early 1920s is responsible for the emergence of ecological studies in sociological research. Place-based theories fall squarely within the theoretical tradition of social ecology, but are more specific about the mechanisms by which structural context is translated into individual action. The dominant theoretical perspectives derive from the routine activities theory (Cohen and Felson, 1979) and rational choice theory (Clarke and Cornish, 1985). In both cases, the distribution of crime is determined by the intersection in time and space of suitable targets and motivated offenders (Luc et al., 2000).

The principal focus of attention of the Chicago School and of debate concerning the ecology of crime, however, was the geographical distribution of the residences of offenders rather than the locations at which crime occurred (Weisburd et al. 2012). From the end of the Second World War to the 1970s, criminology privileged person over place. The potential for geography to gain greater prominence in crime analysis thereafter improved, with the emergence of economic perspectives on crime in the form of rational choice theory (Clarke and Cornish, 1985). Collectively, these opportunity theories of crime ushered in a rich new set of ideas with which to explore criminal activity, but with the focus firmly on micro-social reasoning, how individuals interact in specific locational contexts (Bannister et al., 2019).

Contemporary theories of crime, place, and space include:

- Defensible space theory, which examines how the design of physical space is related to crime.
- Broken windows theory, which looks at the relationship between low-level disorder and crime.
- Routine activities theory, which considers how opportunities to commit crime are shaped by people's everyday movements through space and time.

Strain theories state that certain strains or stressors increase the likelihood of crime. Crime may be a way to reduce or escape from strains. Crime may be used to seek revenge against the source of strain or related targets. Merton developed the first major strain theory of crime in the 1930s. Strain theories are among the dominant explanations of crime (Deflem, 2003). Beginning in the 1960s and 1970s, criminologists began to suggest that the inability to achieve monetary success or middle-class status was not the only important type of strain. They claimed that the inability to achieve any of these goals might result in delinquency. So, strain theories are based on a simple, commonsense idea: When people are treated badly, they may become upset and engage in crime (Agnew, 2008). Subcultural theories build upon the work of Merton. According to Cohen, lower-working-class population want to achieve the success which is valued by mainstream culture. This results in status frustration, as the person is at the bottom of the social structure and have little chance of gaining a higher status in society. In these subcultures the individual who lacked respect in mainstream society can gain it by committing crimes such as vandalism and truancy.

Social control theory has remained a major paradigm in criminology since its introduction in 1969. According to Hirschi (1969), virtually all existing criminological theories began with a faulty fundamental premise: that criminal behavior requires, in some form, the creation of criminal motivation. As Walklate observes, this theory lends itself to the range of policy initiatives known as situational crime prevention, sometimes referred to as designing out crime. This is the umbrella term for a range of strategies that are used to reduce the opportunities to commit crime. Left realism is a main school of thought within critical criminology. While left realism may not be as popular in critical criminological circles as it was in the 1980s and early 1990s, it is still at the forefront of an unknown number of progressive criminologists' minds (Walter, 2010).

Left realists also support two other key theories to explain crime:

- Marginalization: Some groups experience marginalization and at different levels (social, political and economic). These groups are on the periphery of society. Lacking political representation, these groups represent themselves and their ways of taking political action include the commission of crime and violence.
- Sub-cultures: Marginalized individuals and groups may come into contact with others who share these experiences, and who then may form their own sub-cultures in which crime and violence may feature.

2.4 THE SPACE AND TIME OF OFFENCE

The spatio-temporal dimensions of crime analysis is the basic need to draw the patterns of crime in a particular region. Crime is not just a simple event which took place in space but it has its temporal considerations also. It is equally important to study the temporal patterns as the analysis of spatial patterns of crime in a region, because it provides opportunities for crime as spatial accessibility. Certain locations are more vulnerable and provide ample opportunities to criminals for crimes. Explanations for this are grounded in Routine Activities Theory (Cohen and Felson, 1979) and Crime Pattern Theory (Brantingham and Brantingham, 1981). In simple terms, the occurrence of a crime requires the juxtaposition of motivated offenders and suitable targets, a situation constrained in time and space. These constraints are defined by the offenders and victims using of time and space, as their activities are bounded by the need to eat, sleep, work, or recreational activity. Moreover, these activities can only occur at a finite number of locations and times; and, that the movement of offenders and victims is not compulsive, but structured, regulated by the daily routines of offenders and victims, and the social and physical environments within which they interact (Brantingham and Brantingham, 2013). Indeed, "a limited number of sites, times and situations constitute the space-time loci for the vast majority of offenses."

2.4.1 Distance Decay

The observed phenomenon of diminishing spatial interaction, as distances grow longer from places familiar, has been named Distance Decay (Manne, 2007). As the distance between offender and victim increase the possibilities of occurring of crime reduces in the proportional way. A number of factors may be responsible for this trend as more time and resources will be required to target distant victims. The frequency also diminishes as the distance increases. There is always likely to be some distance between home and where a crime is committed. The availability of financial resources of an individual affect both the choice of a target and the methods of travel to that target.

2.4.2 Distribution of Crime Events

The past two decades have seen a major expansion into the analysis of the spatial distribution of crime, with small-scale or micro-level analysis emerging at the forefront of place-based research (Sherman al., 1989). At present, criminology of place has returned geography to the center stage, cloaked in the micro-localities of environmental criminology. This trend has been driven by both the increased availability of spatially referenced crime data and the technological advances of software products which promote the analysis of the spatial clustering of crime, or hot-spot analysis. However, this growth in spatial analysis is perhaps not reflected by similar advances in the temporal analysis of crime. Whilst a number of studies have examined the temporal patterns of crime (Ashby and Bowers, 2013). These theories are not as prominent in the field as the spatial literature. As highlighted over 10 years ago, whilst the spatial analysis of crime has thrived, analysis of the temporal distribution of crime has failed to keep pace (Ratcliffe, 2002). The criminologist should incorporates meaningful multi-level analysis of place and multi-period dynamic causation of crime to root out the cause and patterns of crime. This is still true today; "the majority of studies linking potentially criminogenic places to elevated levels of crime across geographical units have been a temporal". As a consequence of this and perhaps compounded by the challenges of employing complex spatio-temporal analysis methods (Ratcliffe, 2010), the inextricable link between space and time is often omitted from place-based or temporal-based crime research (Ratcliffe, 2010). With the exceptions perhaps of the near repeat victimization literature (Johnson et al., 2007), animated visualizations of sequences of hot spots over the course of the day (Brunsdon et al., 2007; Townsley, 2009) and some isolated studies now discussed, there is a paucity of research into the patterns and manifestations of crime events in both space and time.

2.4.3 Spatial-Temporal Crime Analysis

Crime analysis involves human tendencies which may have not a uniform pattern in different geographical regions. Like other human activities, distribution of crime events and spots may have more concentration in some areas while less in some others. Several researchers acting independently, using data on different crimes and from different nations, have found that crime hot spots shift quickly in response to the structure of daily life. For example, major shifts have been found in robbery locations from afternoon to early morning, and weekday to weekend within the vicinity of schools, parks, and late night businesses (Adams et al., 2015). Others have found high crime risk in some entertainment districts in the early evening, while other entertainment districts experience more crime problems after midnight. Crime near bars and pubs is significant on weekends, but such clustering may be barely noticeable on weekdays (Newton and Hirschfield, 2009). It is very essential to highlight the major concentration areas of crime in an area so as to prepare the rescue and other measures. One such technique is hot spot plotting method. It is a visualization method which aims to present spatial analysis with consideration of the distribution of events in time within hot spots. Another method which is widely used by researchers to draw patterns of crime is Kulldorff's Space-Time Scan Statistic Method. The spatial-temporal scan statistic is based on the spatial scan statistic. The spatial scan statistic is viewed as a 2D crime map, which uses a circular window scanning the study area. While after adding a time factor the spatial-temporal scan statistic employs a 3D cylinder to scan the area both horizontally and vertically. The spatial-temporal scan statistic is used to detect crime clusters in space and time (Shuzhan, 2014).

Crime on transit systems has been shown to be highly dynamic and related to surrounding environments with distinct patterns in both space and in time (Newton et al., 2014). Shiode et al. (2015) found that within high crime areas in Chicago, different micro-scale spatio-temporal crime patterns were evident for different types of crime; drugs, robbery, burglary, and vehicle crime all had their own unique spatial-temporal crime patterns. Haberman and Ratcliffe suggest that the criminogenic nature of places is influenced by a number of factors that include; the length of time facilities are open and the consistency of use during the day. For example, facilities with a steady flow of people versus those with concentrations of people at peaks and sparse use at off peak times; and unofficial use of places when they are in effect closed or recently closed.

Many of the old ideas from Chicago in the 1930s and 1940s no longer hold. Areas identified as high crime parts of towns and cities experience low and moderate levels of crime during certain time periods across several of their streets/blocks. Some areas are prone to certain crime offenses at particular times of the day, but rarely does analysis consider whether these areas suffer from other crime types, either simultaneously or at another time or day of the week. Moreover, little attention is accorded to explaining the dynamics of crime hot spots. Crime can shift rapidly over the course of a 168 hours period (1 week). Furthermore, and especially when there is mixed land-use, characteristics of these populations are likely to differ substantially from the residential population, making it difficult to calculate realistic crime rates.

2.4.4 Understanding Crime in Time and Space

There is a potential crime generator area in almost every urban habitat in the world with a high concentration of people and vulnerable objects, where public gathering becomes the subject matter of criminal activity. This potential area or region attracts the potential offenders from nearby places as well as criminal and offenders from distant places. Crime attractors are objects, areas, settlements, or districts where a high number of (potential) offenders are drawn to for criminal behavior. These targets also form the nodes of the activity of repeat offenders. They strongly and directly influence crime behavior. Typical attractors are catering services with alcohol outlets, drug trafficking places, and entertainment areas of nightlife, but also large shopping malls, especially those located near transport hubs and unguarded parking areas (Natalia and Michael, 2017). Andresen and Malleson explore spatial-temporal patterns by time of day and day of week. Tompson and Bowers scrutinize how weather and seasonality influence the time and location of street robbery; and Malleson and Andresen address the issue of crime risk through the use of the ambient population. The underlying population at risk is itself dynamic, changing in time and place, and is not well represented through use of residential population as a crime denominator.

Andresen and Malleson explicitly explored how the day of week impacts on the spatial and temporal patterns of crime offenses for the city of Vancouver, Canada. They investigate the intra-week patterns of a range of crime types and found increased levels of crime on weekends in certain localities. For example, on Saturdays, theft from vehicles increased in the downtown parks and recreational park areas, and assaults also increased in the bar districts. However, not all crime types revealed expected intra-week patterns. For example, increases in burglary in particular places were observed on Mondays. For robbery and sexual assault, they did not find unique intra-week patterns. This may be due to different groups of offenders operating on different days. However, for most crime types examined there were distinctive temporal and spatial patterns observed for different days of the week.

Tompson and Bowers examine the impact of weather on the spatial-temporal patterns of street robbery. They tested two hypotheses. The first of these is that people's use of space would be influenced by extremes in weather; for example excess heat and extreme cold might limit the use of outdoor space an essential component of street robbery whereas unexpectedly mild or favorable weather might encourage people to venture outside. They found that wind speed and temperature did affect robbery, the adverse impact of winter corresponding with a reduction in robbery, whereas an increase in temperature led to more robberies. However, these variables interacted, as despite increased temperature an increased wind speed in summer months resulted in a decrease in robberies. Thus both variables contributed to what the authors term a person's sense of thermal comfort. The authors move beyond this with their second hypothesis, to examine how weather might impact on discretionary activities, those a person pursues through choice, as opposed to obligatory routine activities they have to do. The hypothesis here is weather will influence the spatial-temporal patterns of discretionary activities more than that of obligatory ones. Temperature, wind speed, and humidity were significant predictors of robbery during the night shift and on weekends, and rain was shown to have a negative relationship with robbery on the weekends. When travel behavior is optional, people are less likely to venture outdoors when it is raining. Thus, weather exerts significant constraints on the space-time loci of crime opportunities, particularly outside of working hours during a person's time and space delineated discretionary routine activities.

Crime researchers and practitioners have put a lot of effort into studying how crime hot spot mapping can be used to assist police decision makers with allocating their limited resources and manpower to areas where crime events are most likely to occur (Shuzhan, 2014). Malleson and Andresen pose a different problem. A key component in the analysis of crime is identifying levels of risk, and crime rates are often used here. For example, identification of burglary risk on a street should take account the number of properties. Violence at night time should consider the number of persons present in the night-time economy. Risk of assault on a train will depend on the number of passengers. Thus, the denominators of crime (crime rates) are an essential component to aid our examination of crime risk. However, when considering crime in both place and time, it is problematic to identify crime rates. Calculating accurately the true population at risk in any given place and time is vital for identification of reliable crime rates. However, data on the movement of persons through areas is not routinely collected during the course of the day, and not simple to capture. Residential populations identified through census and other surveys do not accurately reflect populations in business centers during the daytime, or residential areas when most people are out at work. Thus, criminal research has confirmed that there are clear patterns to crime, with concentrations in specific places and at specific times (Cozens, 2007a).

Concluding Remarks

Spatial crime analysis has become an integral part of geographical research, keeping in view the data used and affected population in mind. As the frequency and intensity of criminal activities in urban areas have increased in a tremendous way, it becomes essential for the researchers and policy makers to develop new tools and techniques in spatial crime mapping and analysis. With the help of new methods as crime hot spot mapping, geographic profiling, distance decay, and Space-Time Scan Statistic Method, the prediction of crime is possible in some extent and curative measures may be adopted on time. The spatial crime mapping by the researchers may be used by law enforcement agencies and may assist in regions which have limited resources and manpower for patrolling and other purposes.

REFERENCES

Adams, W., Herrmann, C., and Felson, M. (2015) Crime, transportation and malignant mixes. In Ceccato, V. and Newton, A. (Eds.), Safety and security in transit environments: An interdisciplinary perspective (pp. 181–195), Palgrave McMillan, Basingstoke, Hampshire. In Newton, A. and Felson, M. (2015) Editorial: Crime patterns in crime and space: the dynamics of crime opportunities in urban areas. Crime Science A Springer Open Journal. https://link.springer.com/content/pdf/10.1186/s40163-015-0025-6.pdf.

Agnew, R. (2008) Strain theory. In Perrillo V. (Ed.) *Encyclopedia of Social Problems*, Sage Publications. https://studysites.sagepub.com/haganintrocrim8e/study/chapter/handbooks/42347_7.1.

Ander, M. A. and Tarah, H. (2018) Predicting property crime risk: An application of risk terrain modelling in Vancouver, Canada. European Journal of Criminal Policy and Research, In Ried S. E., Tita G., and Valasik M. (2019) *The Mapping and Spatial Analysis of Crime*. https://researchgate.net/publication.

Andreza, A de S. S. (2009) Urban Planning, Social Cohesion and Safety in the City of Brasilia, Thesis.

Andresen, M. A. (2006) Crime measures and the spatial analysis of criminal activity. British Journal of Criminology, 46, 258–285. In Andresen, M. A. (2015) *Spatial Dynamics and Crime*. The International Encyclopedia of the Social and Behavioral Sciences, 2nd ed. https://www.researchgate.net/publication/304193503.

Andresen, M. A. (2011) The ambient population and crime analysis. Professional Geographer, 63, 193–212. In Andresen, M. A. (2015) *Spatial Dynamics and Crime*. The International Encyclopedia of the Social and Behavioral Sciences, 2nd ed. https://www.researchgate.net/publication/304193503.

Andresen, M. A. (2013) Science of crime measurement: Issues for spatially-referenced crime data. In Andresen, M. A. (2015) *Spatial Dynamics and Crime*. The International

Encyclopedia of the Social and Behavioral Sciences, 2nd ed. Routledge, New York NY. https://www.researchgate.net/publication/304193503.

Andresen, M. A. and Jenion, G. W. (2010) Ambient populations and the calculation of crime rates and risk. *Sec. J.*, 23, 114–133. In Andresen, M. A. (2015) *Spatial Dynamics and Crime*. The International Encyclopedia of the Social and Behavioral Sciences, 2nd ed. https://www.researchgate.net/publication/304193503.

Andresen, M. A. and Linning, S. J. (2012) The (in)appropriateness of aggregating across crime types. *Appl. Geog.*, 35, 275–282. In Andresen, M. A. (2015) *Spatial Dynamics and Crime*. The International Encyclopedia of the Social and Behavioral Sciences, 2nd ed. https://www.researchgate.net/publication/304193503.

Andresen, M. A. and Malleson, N. (2011) Testing the stability of crime patterns: Implications for theory and policy. *J. Res. Crime Delin.*, 48, 58–82. In Andresen, M. A. (2015) *Spatial Dynamics and Crime*. The International Encyclopedia of the Social and Behavioral Sciences, 2nd ed. https://www.researchgate.net/publication/304193503.

Anderson, M. A. and Tarah, H. (2018) Predicting property crime risk: An application of risk terrain modeling in Vancouver, Canada. *Eur. J. Crim. Pol. Res.*, 24, 373–392. 10.1007/s10610-018-9386-1.

Andrew, N. and Marcus, F. (2015) Editorial: Crime patterns in time and space: The dynamics of crime opportunities in urban areas. *Crime Sci.*, 4(11). 10.1186/s40163-015-0025-6.

Anselin, L., Cohen, J., Cook, D., Gorr, W., and Tita, G. (2000) Spatial analysis of crime. *Crim. Justice*, 4. https://dds.cepal.org/infancia/guia-para-estimar-la-pobreza-infantil/bibliografia/capitulo-IV/Anselin.

Bannister, J., O'Sullivan, A., and Ellie, B. (2019) Place and time in the criminology of place. *Theoretic. Criminol.*, 23(3), 315–332. https://journals.sagepub.com/doi/pdf/10.1177.

Beccalossi, C., Lombroso, C. (2010) The criminal man. In Cullen, F. and Wilcox, P. (Eds.) *Encyclopedia of Criminological Theory*, Sage Publication. studysites.sagepub.com/schram/study/materials/reference/90851_04.1r.pdf.

Boggs, S. L. (1965) Urban crime patterns. *Am. Sociol. Rev.*, 30, 899–908. In Andresen, M. A. (2015) *Spatial dynamics and crime*. The International Encyclopedia of the Social and Behavioral Sciences, 2nd ed. https://www.researchgate.net/publication/304193503.

Bottoms, A. E. (2007) Place, space, crime and disorder. In Maguire, M., Morgan, R., and Reiner, R. *Oxford Handbook of Criminology*, 4th ed. Oxford University Press, Oxford. http://cw.routledge.com/textbooks/9780415464512.

Brantingham, P. L. and Brantingham, P. J. (1981) Notes of the Geometry of Crime. In Brantingham, P. J. and Brantingham, P. L. (Eds.), *Environmental Criminology*, Waveland Press, Prospect Heights, IL, 27–54. In Andresen, M. A. (2015) *Spatial Dynamics and Crime*. The International Encyclopedia of the Social and Behavioral Sciences, 2nd edn. https://www.researchgate.net/publication/304193503.

Brown, R. O., Rossmo, D. K., Sisak, T., Trahern, R., Jarret, J., and Hanson, J. (2005) Geographic Profiling Military Capabilities. Final Report Submitted to the Topographic Engineering Center, Department of the Army, Fort Belvoir, VA. In Andresen, M. A. (2015) *Spatial Dynamics and Crime*. The International Encyclopedia of the Social and Behavioral Sciences, 2nd ed. https://www.researchgate.net/publication/304193503.

Brunsdon, C., Corcoran, J., and Higgs, G. (2007) Visualising space and time in crime patterns: A comparison of methods. *Comp., Environ. Urban Syst.*, 31(1), 52–75.

Canter, D. and Alison, L. (Eds.). (2000) Profiling Property Crimes. Dartmouth Publishing Company Limited, Aldershot, England. Burlington, U.S.A., Ashgate Publishing Limited. In Manne Laukkanen (2007) Geographic Profiling: Using Home to Crime Distances and Crime Features to Predict Offender Home Location, Academic Dissertation. https://www.researchgate.net/publication/274697121.

Caplan, J. M. and Kennedy, L. W. (2016) Risk terrain modeling: Crime prediction and risk reduction. University of California Press, Berkeley. In Vildosola, D., Carter, J., Louderback, E. R., and Roy, S. S. (2019) *Crime in An Affluent City: Applications of Risk Terrain Modelling for Residential and Vehicle Burglary in Coral Gables*, Florida. Applied Spatial Analysis and Policy. https://link.springer.com/article/10.1007/s12061-019-09311-9.

Caplan, J. M. and Leslie, W. K. (2016) Risk terrain modelling: Crime prediction and risk reduction. In Ried S. E., Tita G., and Valasik, M. (2019) *The Mapping and Spatial Analysis of Crime*. University of California Press, Berkeley, CA. https://researchgate.net/publication.

Caplan, J. M., Kennedy, L.W., and Miller, J. (2011) Risk terrain modeling: Brokering criminological theory and GIS methods for crime forecasting. *Justice Quarterly*, 28(2), 360–381. In Vildosola, D., Carter, J., Louderback, E. R. and Roy, S. S. (2019) Crime in An Affluent City: Applications of Risk Terrain Modelling for Residential and Vehicle Burglary in Coral Gables, Florida. *Appl. Spat. Analy. Pol.* https://link.springer.com/article/10.1007/s12061-019-09311-9.

Carcach, C. and Muscat, G. (2002) Location Quotients of Crime and Their Use in the Study of Area Crime Careers and Regional Crime Structures. *Crime Prev. Community Saf.* In Natalia Sypion-Dutkowska and Michael Leitner (2017) *Land Use Influencing the Spatial Distribution of Urban Crime: A Case Study of Szczecin*, Poland. International Journal of Geo-Information.

Chockalingham, K. (2003) Criminal Victimization in Four Major Cities in Southern India. *Forum Crime Soc.*, 3, (1 and 2), United Nations Office on Drugs and Crime.

Clarke, R. V. G. and Cornish, D. B. (1985) Modeling offenders' decisions: A framework for research and policy. *Crime and Justice*, 6, 147–185. In Andresen, M. A. (2015) *Spatial Dynamics and crime*. The International Encyclopedia of the Social and Behavioral Sciences, 2nd ed. https://www.researchgate.net/publication/304193503.

Cohen, L. E. and Felson, M. (1979) Social change and crime rate trends: A routine activity approach. *Am. Sociol. Rev.*, 44, 588–608. In Andresen, M. A. (2015) *Spatial Dynamics and Crime*. The International Encyclopedia of the Social and Behavioral Sciences, 2nd ed. https://www.researchgate.net/publication/304193503.

Davis, M. (1999) *Ecology of Fear: Los Angeles and the Imagination of Disaster*. New York, Vintage Books. http://cw.routledge.com/textbooks/9780415464512.

De Keseredy, W. S. (2010) Left Realism Criminology. In Cullen, F. T. and Wilcox, P. (Eds.), *Encyclopedia of Criminological Theory*, SAGE Publications (online). http://dx.doi.org/10.4135/9781412959193.n150.

Esteves, A. (1995) A Criminalidade Urbana e a Percepção do Espaçonacidade de Lisboa: Uma Geografi a da Insegurança, Faculdade de Letras da Universidade de Lisboa. In Santana, P. et al. (2013) *Crime: Impacts of Urban Design and Environment*. https://www.researchgate.net/publication/307766341.

Federal Bureau of Investigation. (2012) Uniform Crime Reports. http://www.fbi.gov/aboutus/cjis/ucr/ucr. In Andresen, M. A. (2015) *Spatial Dynamics and Crime*. The International Encyclopedia of the Social and Behavioral Sciences, 2nd ed. https://www.researchgate.net/publication/304193503.

Ferreira, E. (1998) Crime e Insegurançaem Portugal: Padrões e Tendências, 1985–1996, Celta Editora, Lisboa. In Santana, P. et al. (2013) Crime: Impacts of Urban Design and Environment: https://www.researchgate.net/publication/307766341.

Fyfe, N. (2000) Crime, Geography of. In *Dictionary of Human Geography*, 4th ed. Edited by Johnston R., Gregory D., Pratt G., and Watts M., 120–123. Oxford, Blackwell. In Yarwood, R. (2015) Geography of Crime. https://www.researchgate.net/publication/328138626.

Guerry, A. M. (1833) Essai Sur la Statistique Morale de la France. In Andresen, M. A. (2015) *Spatial Dynamics and Crime*. The International Encyclopedia of the Social and

Behavioral Sciences, 2nd ed. Crochard, Paris. https://www.researchgate.net/publication/304193503.

Herbert, S. and Brown, E. (2006) Conceptions of space and crime in the punitive neoliberal city. Antipode, 38(4). Wiley Online Library. https://onlinelibrary.wiley.com/doi/abs/10.1111/j.1467-8330.2006.00475.x.

Hirschi, T. (1969) Causes of delinquency. University of California Press, Berkeley. http://pgil.pk/wp-content/uploads/2014/04/KEY-IDEAHIRSCHI%E2%80%99S-SOCIAL.pdf.

Johnson, S. D., Bernasco, W., Bowers, K. J., Elffers, H., Ratcliffe, J., Rengert, G., and Townsley, M. (2007) Space-time patterns of risk: A cross national assessment of residential burglary victimization, *J. Quanti. Criminol.* https://link.springer.com/article/10.1007/s10940-007-9025-3.

Jerry, H. R. (2014) Crime Mapping: Spatial and Temporal Challenges. https://www.researchgate.net/publication/226797328.https://www.researchgate.net/publication/304193503.

Lawman, J. (1982) Conceptual Issues in Geography of Crime. Toward a Geography of Control Annals, Association of American Geographers, 76, pp 81–94. In Ahmadi, M. (2003) Crime Mapping and Spatial Analysis. Thesis Submitted to the International Institute for Geo-information Science and Earth Observation. https://pdfs.semanticscholar.org/26fc/afccd32e738286a20194dace82691c80ff0a.pdf.

Luc, A., Jacqueline, C., David, C., Wilpen, G., and George, T. (2000) Spatial Analyses of Crime, Measurement and Analysis of Crime and Justice.

Lucy, W. M. and Alexander, Z. (2014) spatial approach to surveying crime-problematic areas at the street level. In Huerta, Schade, Granell (Eds), Connecting a Digital Europe through Location and Place. Proceedings of the AGILE'2014 International Conference on Geographic Information Science, Castellón, June, 3–6, 2014. ISBN: 978-90-816960-4-3.

Manne, L. (2007) Geographic Profiling: Using Home to Crime Distances and Crime Features to Predict Offender Home Location, Academic Dissertation. https://www.researchgate.net/publication/274697121.

Mark, S., Jan, V. D. and Wolfgang, R. (2003) Determining Trends In Global Crime And Justice: An Overview of Results from The United Nations Surveys of Crime Trends and Operations of Criminal Justice Systems. In Forum On Crime And Society, 3 (1 and 2), United Nations Office on Drugs and Crime.

Martin, A. A. (2017) *Spatial Dynamics and Crime*, The International Encyclopaedia of the Social and Behavioural Sciences, 2nd ed.

Mosher, C. J., Miethe, T. J., and Hart, T. C. (2011) The mismeasure of crime. In Andresen, M. A. (2015) *Spatial Dynamics and Crime*. The International Encyclopedia of the Social and Behavioral Sciences, 2nd edn, Sage Publications, Los Angeles, CA. https://www.researchgate.net/publication/304193503.

Natalia, S. D. and Michael, L. (2017) Land use influencing the spatial distribution of urban crime: A case study of Szczecin, Poland. ISPRS *Inter. J. Geo-Inform.*, 6(3), 74. https://www.mdpi.com/2220-9964/6/3/74.

Newman, O. (1972) Defensible space. In Santana, P., Santos, R., Costa, C. and Loureiro, A. (2013) *Crime: Impacts of Urban Design and Environment* New York, Macmillan. https://www.researchgate.net/publication/307766341.

Newton, A. and Hirschfield, A. (2009) Measuring violence in and around licensed premises: The need for a better evidence base. *Crime Prev. Comm. Saf.*, 11(3), 171–188. In Newton, A. and Felson, M. (2015) Editorial: Crime patterns in crime and space: the dynamics of crime opportunities in urban areas. Crime Science A Springer Open Journal. https://link.springer.com/content/pdf/10.1186/s40163-015-0025-6.pdf.

Newton, A., Partridge, H., and Gill, A. (2014) Above and below: Measuring crime risk in and around underground mass transit systems. *Crime Sci.*, 3(1), 1–14. In Newton, A. and Felson, M. (2015) Editorial: Crime patterns in crime and space: The dynamics of crime opportunities in urban areas. Crime Science A Springer Open Journal. https://link.springer.com/content/pdf/10.1186/s40163-015-0025-6.pdf.

Perreault, S. and Brennan, S. (2010) Criminal Victimization in Canada, 2009. In Andresen, M. A. (2015) *Spatial Dynamics and Crime*. The International Encyclopedia of the Social and Behavioral Sciences, 2nd ed. Ottawa, Statistics Canada.

Quételet, L. A. J. (1842) A treatise on man. In Andresen, M. A. (2015) *Spatial Dynamics and Crime*. The International Encyclopedia of the Social and Behavioral Sciences. 2nd ed. Edinburgh, William and Robert Chambers. https://www.researchgate.net/publication/30419350.

Ratcliffe, J. H. (2002) Aoristic signatures and the spatio-temporal analysis of high volume crime patterns. *J. Quant. Criminol.*, 18(1), 23–43. In Newton, A. and Felson, M. (2015) Editorial: Crime patterns in crime and space: The dynamics of crime opportunities in urban areas. Crime Science A Springer Open Journal. https://link.springer.com/content/pdf/10.1186/s40163-015-0025-6.pdf.

Ratcliffe, J. H. (2010) Crime mapping: Spatial and temporal challenges. In Piquero, A. R. and Weisburd, D. (Eds.), *Handbook of Quantitative Criminology* (pp. 5–24), Springer, New York, NY. In Newton, A. and Felson, M. (2015) Editorial: Crime patterns in crime and space: The dynamics of crime opportunities in urban areas. Crime Science A Springer Open Journal. https://link.springer.com/content/pdf/10.1186/s40163-015-0025-6.pdf.

Richard, Y. (2015) Geography of Crime. https://www.researchgate.net/publication/328138626.

Rossmo, D. K. (2012) Recent developments in geographic profiling., 6, 144–150. In Andresen, M. A. (2015) *Spatial Dynamics and Crime*. The International Encyclopedia of the Social and Behavioral Sciences, 2nd ed. https://www.researchgate.net/publication/304193503.

Rossmo, D. K. (2014) Geographic profiling. In Weisburd, D., Bruinsma, G. J. D. (Eds.), *Encyclopedia of Criminology and Criminal Justice*. Springer-Verlag, New York, NY, in press. In Andresen, M. A. (2015) *Spatial Dynamics and Crime*. The International Encyclopedia of the Social and Behavioral Sciences, 2nd ed. https://www.researchgate.net/publication/304193503.

Rossmo, D. K. and Harries, K. D. (2011) The geospatial structure of terrorist cells. *Justice Quart.*, 28, 221–248. In Andresen, M. A. (2015) *Spatial Dynamics and Crime*. The International Encyclopedia of the Social and Behavioral Sciences, 2nd ed. https://www.researchgate.net/publication/304193503.

Sam, J. C. (2019) *Social and Physical Neighborhood Effects and Crime: Bringing Domains Together Through Collective Efficacy Theory*. MDPI, Basel, Switzerland.

Sherman, L. W., Gartin, P., and Buerger, M. E. (1989) Hot spots of predatory crime: Routine activities and the criminology of place. *Criminology*, 27, 27–55. In Andresen, M. A. (2015) *Spatial Dynamics and Crime*. The International Encyclopedia of the Social and Behavioral Sciences, 2nd ed. https://www.researchgate.net/publication/304193503.

Shiode, S., Shiode, N., Block, R., and Block, C. (2015) Space-time characteristics of micro-scale crime occurrences: An application of a network-based space-time search window technique for crime incidents in Chicago. *Internat. Encyclo. Soc. Behav. Sci.* http://www.tandfonline.com/doi/abs/10.1080/13658816.2014.968782?journalCode=tgis20. In Newton, A. and Felson, M. (2015) Editorial: Crime patterns in crime and space: The dynamics of crime opportunities in urban areas. Crime Science A Springer Open Journal. https://link.springer.com/content/pdf/10.1186/s40163-015-0025-6.pdf.

Shuzhan, F. (2014) The Spatial-Temporal Prediction of Various Crime Types in Houston, TX Based on Hot-Spot Techniques. Thesis Submitted to the Graduate Faculty of the Louisiana State University and Agricultural and Mechanical College. https://digitalcommons.lsu.edu/gradschool_theses.

Smith, W. R., Frazee, S. G., and Davison, E. L. (2000) Furthering the integration of routine activity and social disorganization theories: Small units of analysis and the study of street robbery as a diffusion process. *Criminology*, 38, 489–523. In Andresen, M. A. (2015) *Spatial Dynamics and Crime*. The International Encyclopedia of the Social and Behavioral Sciences, 2nd ed. https://www.researchgate.net/publication/304193503.

Townsley, M. (2009) *Spatial Autocorrelation and Impacts on Criminology, Geographical Analysis*. Wiley-Blackwell Publishing. https://research-repository.griffith.edu.au/bitstream/handle/10072/29701/60434_1.pdf?sequence=2.

The Chicago School and Cultural/Subcultural Theories of Crime. agepub.com/sites/default/files/upm-binaries/29411_6.pdf. https://revisesociology.wordpress.com/2012/05/12/3-subcultural-theories/.

UN-HABITAT. (2007) Enhancing Urban Safety and Security, Global Report on Human Settlements 2007, United Nations Human Settlements Programme. https://www.Un.Org/Ruleoflaw/Files/Urbansafetyandsecurity.

Vann, I. B. and Garson, G. D. (2001) Crime mapping and its extension to social science analysis. *Soc. Sci. Comp. Rev.*, 19(4), 471–479. https://www.researchgate.net/publication/328138626.

Weisburd, D., Bushway, S., Lum, C., and Yang, S.-M. (2004) Trajectories of crime at places: A longitudinal study of street segments in the City of Seattle. Criminology, 42, 283–321. In Andresen, M.A. (2015) *Spatial Dynamics and Crime*. The International Encyclopedia of the Social and Behavioral Sciences, 2nd edn. https://www.researchgate.net/publication/304193503.

Weisburd, D., Groff, E., and Yang, S. M. (2012) *The Criminology of Place: Street Segments and Our Understanding of the Crime Problem*. Oxford University Press, New York. In Bannister, J., O'Sullivan, A. and Bates, E. (2017) *Place and Time in the Criminology of Space*. Theoretical Criminology, Vol. 23, Sage. 10.1177/1362480617733726.

Weisburd, D., Groff, E. R., and Yang, S.-M. (2013) Understanding and controlling hot spots of crime: The importance of formal and informal social controls. Prevention Science, in press. In Andresen, M. A. (2015) Spatial Dynamics and Crime. *Inter. Encyclop. Soc. Behavi. Sci.*, 2nd edn. https://www.researchgate.net/publication/304193503.

White, R. and Sutton, A. (1995) Crime prevention, urban space and social exclusion. *Australi. New Zealand J. Sociol.*, 31(1), 82–99.10.1177/144078339503100106.

3 The Geography of Neighborhood Studies

India's geostrategic location in South Asia, its relatively sound economic position in comparison to its neighbors, and its liberal democratic credentials, have given it a political edge in the region. India shares 15,106.7 km of its boundary with seven nations: Pakistan, China, Nepal, Bhutan, Myanmar, Bangladesh, and Afghanistan. These land borders run through different terrains. Managing a diverse land border is a complex task with a boundary of 7,516.6 km, which includes 5,422.6 km of coastline in the mainland and 2,094 km of coastline bordering islands. The coastline touches nine states and two union territories. The traditional approach to border management (i.e., focusing only on border security), has become inadequate. India needs to not only ensure seamlessness in the legitimate movement of people and goods across its borders but also undertake with the adoption of new technologies for border control and surveillance and the development of integrated systems for entering, exchange and storage of data, will facilitate the movement of people and products without endangering security. An Indian perspective explores how the government of India can respond to border management challenges curbing crime occurrence and adopt a proactive and resilient approach towards smart border management that should have four key elements: innovation and technology infrastructure, collaborative border management, capacity building, and agile organization.

3.1 INFILTRATION, REFUGEE MOVEMENTS, AND CROSS-BORDER TERRORISM

Displacement is a fact of life for many people across the world and the refugee problem is a tragic phenomenon, a by-product, of modern times. It is a product, not only of world war(s), modern dictatorial regimes, and ethnic strife, but also of the general socioeconomic inequalities that rule the world. The 20th century has been described as the "century of the uprooted" as a result of the tremendous increase in the number of refugees, homeless, and displaced people around the world, particularly in the Third-World nations. These people are driven by economic, environmental, political, and other push-and-pull factors, and presence of these people also poses serious threat to the social, economic, and political institutions in the host country. Threats to social security come in the form of drug addiction, drug trafficking, and crimes against women. In the recent past, a new dimension has emerged in terrorism (i.e., transnational Terrorism). It involves incidents where the perpetrators and victims are from two or more countries. Infiltration from across the border has also become a significant threat to all the nations, and every responsible state is trying to secure their tertiary from these threats. There have been numerous instances in the past when criminal organizations, particularly terrorists, use porous borders and wreak havoc on the lives of innocent people in destination countries.

3.1.1 Infiltration

Cross-border infiltration has become a significant threat to all the nations, and every responsible state is trying to secure their tertiary from these threats coming from across the border. In the period after the end of the Cold War, the impact of transnational threats to the national security of most of the South Asian nations has greatly increased. There have been numerous instances in the past when criminal organizations, particularly terrorists, use porous borders and wreak havoc on the lives of innocent people in destination countries. Take the example of Islamist militant groups which remains Pakistan's most successful strategic weapon against India's regional hegemony. Having lost every war against its much larger and conventionally superior neighbor, Pakistan has been fighting a long-running proxy war against India, particularly in the states of Jammu and Kashmir, which have been affected by terrorist violence that is sponsored and supported from across the border. To sneak into Indian territory, the number of infiltration attempts made by Pakistan-based terrorist groups increased from 222 in 2014, to 284 in 2018 (MHA, 2018). Net estimated infiltration from Pakistan to India was 65 during 2014, which has substantially increased to 128 in 2018, the highest in the last 5 years. Infiltration has also caused devastating effects in northeastern states where local population felt "threatened by an increasing sense of being marginalized in their home land by a culturally and ethnically different group". This has become serious issue since the total population is increasing linearly year by year.

It would be difficult and unfair to exclude either Afghanistan or Myanmar from consideration in this perspective. Instability in these countries spills over into the larger south Asian region and leads to the exacerbation of threats posed by drugs, the smuggling of weapons, and terrorism. Likewise, it is difficult to exclude the effect of the spillover of the problems of Afghanistan and Myanmar into Pakistan, India, and Bangladesh, which adds a significant dimension to the national security concerns of these countries. Pakistan claimed that the militant network responsible for most attacks in its territory has their origin from across the border which includes Afghan, Chechen, Arab, Uzbek, and Tajik fighters who find easy access to Pakistan through the porous Afghan-Pakistan border. Most unlawful migrants to Malaysia are from nearby Southeast Asian countries such as Indonesia, particularly on Indonesia-Malaysia border via the maritime boundary of Strait of Malacca and land border on the island of Borneo. The presence of refugees also poses serious threats to the social, economic, and political institutions in the host country. Threats to social security come in the form of drug addiction, drug trafficking, increasing law and order problems, trafficking in weapons and women, etc.

3.1.2 Refugee Movements

Refugees include individuals recognized under the 1951 Convention relating to the Status of Refugees; its 1967 Protocol; the 1969 OAU Convention Governing the Specific Aspects of Refugee Problems in Africa; those recognized in accordance with the UNHCR Statute; individuals granted complimentary forms of protection; or those enjoying temporary protection. The issue of arriving at a competent definition for the term "refugee" is important because refugee status is a privilege or entitlement, given to those who qualify access to certain scarce resources or services outside their own country, such as admission into another country ahead of a long line of claimants, legal protection abroad, and often some material assistance from public or private agencies. According to the definition above, groups who are likely to become refugees are dissidents, target minorities, victims of violence, and victims of massive human rights abuse.

The refugee problem is a tragic phenomenon which has become a matter of acute international concern. It is a by-product of modern dictatorial regimes and ethnic strife as well as general socioeconomic inequalities that rule the world. The 20th century has been described as the "century of the homeless" and the "century of the uprooted" as a result of the tremendous increase in the number of refugees, homeless, and displaced people around the world, particularly in the Third-World nations. During modern times, refugee movements originally started with all the displaced people in Europe after World War II, and subsequently spread to the rest of the world, especially to the Third-World countries. Every movement has given rise to a whole new class of people, who are homeless and stateless, who live in conditions of penury and constant insecurity, who often cause grave political and socioeconomic problems for the host countries. The causes of refugee situations and the reasons which prompt people to flee, sometimes at great physical risk, are fairly evident. Clearly, military action and the instinct of self-preservation are high on the list (James, 1982). International conflicts, revolutions, coups, regime changes which led to political instability and antagonism, ethnic, communal conflicts, and restructuring of state boundaries are among the most prominent motives for flight. Similarly, dissatisfaction with a political system which creates unacceptable economic conditions is also a strong stimulant to the search for a better life elsewhere.

India is geographically situated between the countries of Golden Triangle and Golden Crescent and is a transit point for narcotic drugs produced in these regions to the West. India also produces a considerable amount of licit opium, part of which also finds place in the illicit market in different forms. Illicit drug trade in India centers around five major substances, namely, heroin, hashish, opium, cannibas, and methaqualone. Seizures of cocaine, amphetamine, and LSD are not unknown but are insignificant and rare. Our borders have traditionally been most vulnerable to drug trafficking. In 1996, out of the total quantity of heroin seized in the country, 64% was sourced from the Golden Crescent. We are now witnessing the highest levels of displacement on record. An unprecedented 70.8 million people around the world have been forced from their homes (UNHCR, 2019). Among them are nearly 25.9 million refugees, over half of whom are under the age of 18. Fifty-seven percent of UNHCR refugees came from Syria, Afghanistan, and South Sudan. As fighting continued throughout the year, the Syria situation remained the world's largest refugee crisis, with humanitarian needs and protection risks staggering in scale and severity. This resulted in a context of complex and overlapping displacements. Along with fighting, hunger and disease have also created one of the worst humanitarian crises of our time.

3.1.3 India and Neighboring Countries

There have been several incidences of crime arousal in the border regions of the country. There are various factors which are responsible for refugee movement involving India and its neighboring countries. The most prominent factor which led to mass exodus of population was formation and restructuring of state boundaries after partition of British India into two separate countries (i.e., India and Pakistan). Another important factor was involvement of large number of people in ethnic and secessionist wars where intervention of massive military operations were need of the hour. Soviet military intervention in Afghanistan, emergence of Bangladesh, and struggle for a separate Tamil Eelam in Sri Lanka created millions of refuges in the Indian subcontinent. Lack of economic opportunities in the countries of origin and environmental factors are also responsible for the creation of large numbers of refugees. Often the host country has exacerbated its own problems by creating situations conducive to attract large numbers of refugees from across the borders. Pakistan's supports to the Afghan mujahidin during their struggle against the Soviet occupation of Afghanistan, and India's intervention in Sri Lanka in the 1980s, are examples which produced a climate enabling the Afghans and Tamils to migrate into Pakistan and India, respectively. Efforts have made to curb drug menace, human trafficking, terror activities, infiltration, and illegal cross border movement in these regions. Organized crime in India and South Asia more generally may be characterized by its diversity of activities (from illicit drugs to counterfeit medicinal products), capture of elements of government, and convergence with terrorism or long-standing insurgencies in particular states and border regions. India's Maoist insurgency, for example, which was responsible for hundreds of deaths, also presents a serious criminal and security problem in some states. Combatting this and other insurgency groups absorbs a substantial proportion of funds allocated to modernize the police. The most serious terrorist attack before the "26/11" (2008) event was the 1993 Mumbai bombings, where over 200 people were killed. This attack allegedly involved forces working across the border and within the state.

Violence has always been the key factor in creating refugees, which is evident from a review of following mass movements involving India and its neighboring countries.

3.1.3.1 India-Pakistan Refugee Flows During 1947–1948

As a result of the partition of British India in 1947, India and Pakistan emerged. The partition was based on religion and as a result, the Muslims of North India migrated to the newly created state i.e., Pakistan, and the resultant communal frenzy pushed the Hindu residents of the newly created Pakistan to migrate to the truncated India. This process continued for nearly a decade, though the largest exchange of population took place in 1947 and 1948 only. No precise estimate of the numbers of people involved in these flows is possible, but the combined flows from both directions could not have been much less than some 15 million people. The problems thrown up by the attempts to resettle and rehabilitate them were predictably enormous; however, both countries were able to sorted them out administratively, within their own respective jurisdictions, and also through bilateral negotiations.

3.1.3.2 Exodus of Burmese Indians During 1948–1965

The end of the colonial era generated multiple flows of refugees in the South Asia region. Soon after independence in 1948, Burma, now Myanmar, started the process of nationalizing its administration and public sector institutions. In these institutions, a large number of persons of Indian origin were employed. They were removed from their jobs and sent back to India as refugees since most of them were not granted citizenship. Again in 1962, military rulers of Burma nationalized various economic establishments, depriving many persons of Indian origin of their livelihood. Due to a political stand-off between the Indian government and military rulers of Burma, an estimated 1,500,00 persons of Indian origin, many of them were working as middlemen and money lenders, were sent back to India as refugees.

3.1.3.3 Exodus of Sri Lankan Indians and Tamils Since 1954–1987

The government of Sri Lanka, through newly introduced citizenship acts in 1948–1949, deprived a large number of Indians of their voting rights. As a result, about 1,000,000 persons employed as estate workers became "stateless." As a result, a new category of stateless Indians emerged in Sri Lanka, which neither India, nor Sri Lanka, wanted to own. After negotiations, India agreed to repatriate some 3,38,000 persons of Indian origin from Sri Lanka for resettlement and rehabilitation between 1964 and 1987.

3.1.3.4 Flight of Tibetans to India During 1958–1963

The expansionist policy pursued by the Chinese government resulted in the internal displacement of over one million Tibetan citizens. The Tibetan struggle for autonomy and the Chinese actions to suppress the movement in 1959 forced Dalai Lama, the religious and political leader of the Tibetans, and thousands of his followers to take asylum in India. The exodus continued since then, though the intensity and volume of this flow has varied, depending upon the intensity of the conflict in Tibet. It is estimated that over 1,000,000 Tibetan refugees came to India, Nepal, and Bhutan, mainly because of geographical proximity and a shared Buddhist cultural identity. After 1988–89, Tibet's struggle for autonomy has picked up momentum again, generating more refugees. The question of Tibetan refugees in South Asia will increasingly come under sharp focus as the issue of Tibetan autonomy gains momentum in view of renewed international support.

3.1.3.5 Flight of Tibetans to Nepal During 1959–1989

Although Nepal is home to some 8 million stateless residents; the exact number of refugees is uncertain because Nepal is not a signatory of the 1951 U.N. Convention Relating to the Status of Refugees that ensures the legal status and economic rights of refugees. Nepal is home to over 30,000 refugees, many of whom fled south across the Himalayas following the 1959 Tibetan uprising against Chinese rule in Lhasa, the capital of the Tibet. Tibetan refugees started arriving in the 1950s, but their first major inflow in Nepal occurred after the 1959 Lhasa uprising. Those who arrived before 1989 were issued refugee ID cards and benefited from de facto economic integration; however, more recent arrivals have no legal status and cannot own property, businesses, vehicles, or be employed lawfully.

3.1.3.6 Flight of People from Bangladesh to India in 1971

The conflict between East and West Pakistan in 1971 was a consequence of the more populous East Pakistan's quest for more autonomy as well as political and economic reforms. The 9-month war led to the creation of the state of Bangladesh. This nation-building process precipitated not only political, ethnic, and religious conflicts but also created economic and environmental conditions forcing about 10,000,000 people to become refugees in India. They returned only when Bangladesh emerged as a new, independent nation in December 1971.

As a consequence of the emergence of Bangladesh, another category of refugees was created, that of "Stranded Pakistanis." Some 4,70,000 persons refused to support the struggle for Bangladesh, and continued to proclaim their allegiance to Pakistan while residing in the newly created nation. However, under an agreement which was concluded in 1974, Pakistan accepted about 1,70,000 of these people, popularly known as "Bihar Muslims," whereas about 3,00,000 people remain stranded in Bangladesh awaiting their repatriation to Pakistan.

3.1.3.7 The Flight of Afghans from Afghanistan to Pakistan During 1978–1993

Saur revolution of Afghanistan in 1978 which was followed by Soviet invasion of Afghanistan in 1979 created another huge exodus, with over one million people crossing the border with Pakistan. The flow of refugees was "linked to the intensity of fighting and economic disruption" in Afghanistan. Between 1978 and 1979, over 1,95,000 Afghan refugees had arrived in Pakistan to seek asylum. By the beginning of the 1980s, there were about 5,000 Afghan refugee entering Pakistan every day, contributing to a total of 3 million registered refugees at the peak of the influx, in 1987. Due to cultural differences, most of the refugees wished to return home, and those of the countries traditionally accepting refugees for resettlement.

3.1.3.8 Flight of Burmese Muslims to Bangladesh in 1978

In 1978, as many as 2,00,000 Rohingyas sought refuge in Bangladesh when the Burmese army launched an operation in the Arakan region to check illegal migrants and fight insurgency. An agreement was worked out between Bangladesh and Burma, with the mediation of UNHCR to repatriate these refugees, many of whom also left for Muslim countries in West Asia. To suppress the pro-democracy movement and control the ethnic insurgencies, the Burmese military regime again confronted Rohingyas in 1989–90. Eventually, both of governments reached on an agreement for the repatriation of these refugees.

3.1.3.9 Flight of Chakmas and Hajong to India in 1981

The Chakma conflict is both a religious and ethnic problem in Bangladesh. The Chittagong Hill Tracts saw tribal Chakma leave the area due to religious and ethnic strife caused by Bangladesh's Islamization policy. Buddhists by faith, the Chakmas faced religious persecution in East Pakistan along with the Hajongs, who are Hindus. In addition, the Chakmas and Hajong fled to East Pakistan in 1964–65, since they lost their land to the development of the Kaptai Dam on the Karnaphuli River. Until 1981, about 1,00,000 Chakmas and Hajong eventually sought refuge in India. The Indian government set up relief camps in Arunachal Pradesh and a majority of them continue to live there even after five decades. According to the 2011 census, 47,471 Chakmas live in Arunachal Pradesh alone. The locals and regional political parties opposed re-settling refugees in their land, fearing that it may change the demography of the state and that they may have to share the limited resources available for them. In 2015, the Supreme Court directed the Center to grant citizenship to Chakmas and Hajongs who had migrated from Bangladesh in 1964–69.

3.1.3.10 Exodus of Bhutanese of Nepali Origin to Nepal in 1990–1991

Bhutanese refugees are a group of Nepali language-speaking Bhutanese people. These refugees registered in refugee camps in eastern Nepal during the 1990s as Bhutanese citizens deported from Bhutan during the protest against the Bhutanese state demanding democracy and different state. As Nepal and Bhutan have yet to implement any agreement on repatriation, many Bhutanese refugees have since resettled to North America, Oceania, and Europe under the auspices of the Office of the United Nations High Commissioner for Refugees (UNHCR). Many of them also migrated to areas of West Bengal and Assam in India independently of the UNHCR.

3.1.3.11 From Bangladesh to Assam in India Since 1971

After the Pakistan crackdown in East Bengal, more than one million refugees sought shelter in Assam. Most of them went back after the creation of Bangladesh, but nearly 1,00,000 remained. Even after 1971, large-scale influx continued from the Bangladesh side. Besides Islamic interests, pro-Bangladesh sentiments, unnatural partition, porous borders, devastating floods, population pressure on land, and lack of economic opportunities, have also pushed thousands from Bangladesh towards India's northeast, particularly in Assam (Kumar, 2006). Saikia (2005) estimates the number of illegal migrants in Assam as 1.4 million and 1.1 million during 1971–1991 and 1991–2001, respectively. The demographic transformation of Assam has created apprehension among Assamese that presence of large number of non-Assamese from across the border may lead to the Assamese being reduced to a minority in their own land and consequently to the subordination of their language and culture, loss of control over their economy and politics, and, in the end, the loss of their very identity and individuality as a people. This feeling has imparted a strong emotional content to their movement against illegal migrants since eighties which has led to power struggle across socio-demographic groups, and witnessed a number of armed conflicts between insurgents and the locals.

3.1.4 Refugee Movement and Crime

International flows of people are a distinctive trait of our contemporary globalized world, in as much as international flows of goods, services, and ideas. However, host countries are far more concerned that refugees increase crime, rather than unemployment or taxes and, hence, fear of their involvement in crime is at the center of the public and political debate. With the number of forcefully displaced people hitting a record 70.8 million in 2018, large-scale arrival and prolonged presence of refugees can have various kinds of impacts in the host country, which are far more difficult to measure. The UNHCR notes that "when large numbers of refugees arrive in a country and especially when they are in a destitute situation and do not share ethnic or cultural linkages with the host community there is always a risk that social tensions, conflicts and even violence might arise." Threats to global social security from refugees, according to the experts, assumes greater significance in the forms of human trafficking, drug trafficking, drug addiction, trafficking in arms, violence, robbery, pick pocketing, thefts, vandalism, forgery of official documents, burglaries, rapes, sexual assaults, property, etc. (Fig. 3.1).

According to a study by the United Nations Office on Drugs and Crime, about 2.5 million migrants were smuggled across borders, an operation worth about $5.5 billion to $7 billion in 2016 alone. Neighboring countries shoulder the entire burden of the situation inspite of the fact that these countries often lack sufficient funding to deal with the mass influx of people, leading to a growth in human trafficking and smuggling. Many of the countries absorbing large flows of refugees do not have comprehensive policies to deal with the situation. Refugees are especially vulnerable as they typically move under desperate situations. This creates a situation where transnational crime organizations can come in and take advantage of them through exploitation or trafficking. In most cases, the movement of refugee populations has been based on ongoing violent armed conflicts in the country of origin. Host countries can also be vehicles of spillover violence, if those arriving bring weapons or militant ideologies with them. There are often inherent conflicts among the refugees themselves based on their religious, ideological, and group loyalties. These factions were found to be involved in the conflict against the state, on the one hand, and with each other, on the other hand. To sustain their two-pronged armed conflicts, these groups at times take refuge in other countries with or without the knowledge of the host country. These loyalties resulted in open fighting among the various groups living in the host country. During 1980s, various factions of the Sri Lankan refugees in Tamil Nadu were openly involved in armed fights. In addition, refugees are often accused of gun running and arms smuggling. In the case of Afghan and Sri Lankan Tamil refugees, they possessed weapons without valid licenses and took the law into their own hands, especially with the lack of direct

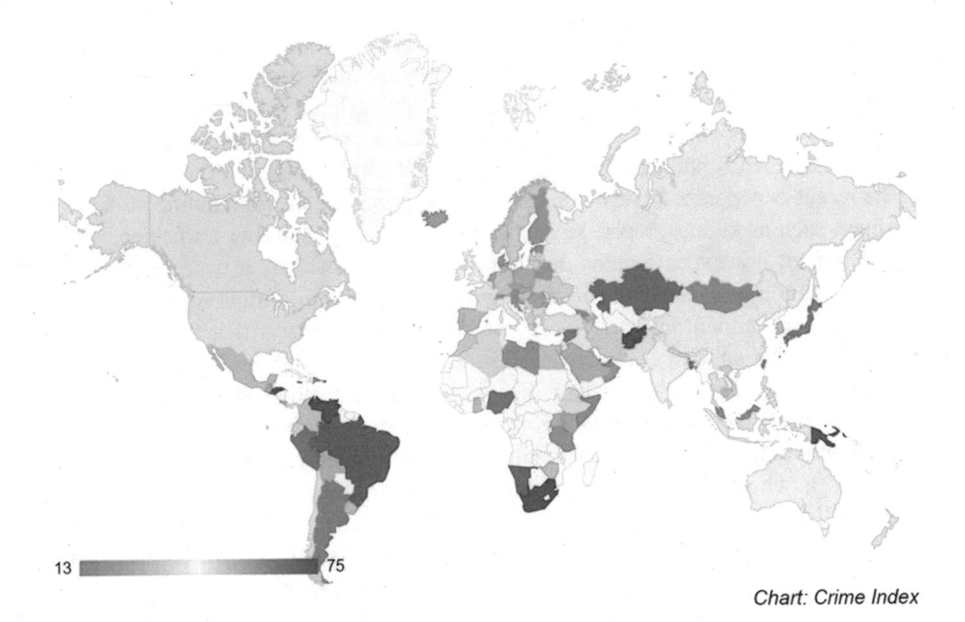

FIGURE 3.1 Global Crime Rate.

control over their activities by the state authorities. The most frequently quoted instance of refugees being a security threat to the state was assassination of Rajiv Gandhi by Sri Lankan Tamils. Review of the literature focusing on recent, high-quality studies in some of the European countries found that various types of criminal activities have taken place in those countries where a large number of refugees have taken shelter. India has a vast coastline of about 7,500 km and open borders with Nepal and Bhutan and is prone to large-scale smuggling of contraband and other consumable items. Though it is not possible to quantify the value of contraband goods smuggled into this country, it is possible to have some idea of the extent of smuggling from the value of contraband seized, even though they may constitute a very small proportion of the actual smuggling.

3.1.5 Terrorism

Terrorism is the premeditated use or threat to use violence by individuals or sub-national groups against noncombatants in order to obtain a political or social objective through the intimidation of a large audience beyond that of the immediate victim. Two essential constituents of the definition are violence and the presence of a political or social motive. Without violence or its threat, terrorists cannot force a political decision maker to respond to their demands. In general, the criteria for defining the term "terrorism" has remained subjective and based mainly on political considerations.

Terrorism has now become intertwined with organized crime, human trafficking, and corruption; no border of the world is untouched by the illicit drug trade today. Terrorism is a serious problem which many countries, including India, are facing. Conceptually, terrorism does not fall in the category of organized crime, as the dominant motive behind terrorism is political and/or ideological and not the acquisition of money-power. The recent experience, however, shows that the criminals are perpetrating all kinds of crimes, such as killings, rapes, kidnappings, gun-running, and drug trafficking.

The fact, however, remains that terrorism is promoted by a wide range of motives depending on the point in time and the prevailing political ideology. Walter Laqueur, an eminent authority on the subject is of the views that there is no definition of terrorism that could cover all its various manifestations in history. Despite differences in approach, most of the experts of the field tend to agree that present-day terrorism is a negative political phenomenon with grave consequences for the individual, society, political regimes, international community, and the human race as a whole. Terrorism can stem from various causes that include: ethno-nationalism, separatism, social injustice, fundamentalist beliefs, religious freedom, etc. In the view of the diverse manifestations of terrorism, it is too arduous to arrive at a common definition of terrorism, which can satisfactorily cover all the varied factors. In other words, terrorism is notoriously difficult to define precisely, objectively, and scientifically except that the primary act of terrorism is to terrorize.

3.1.5.1 Transnational Terrorism

In the recent past, a new dimension has emerged in terrorism (i.e., transnational terrorism). It involves incidents where the perpetrators and victims are from two or more countries. If an incident begins in one country but terminates in another, then the incident is called transnational terrorism. The country can unleash the terrorist activities in a different capacity (i.e., by pertaining the acts of terrorism in the territory of the other country with the purpose of destabilizing that country). They may use their own directly recruited and controlled terror squads, or may choose to work through proxies. These countries invariably work covertly in such support so that they are able to plausibly deny their involvement. This kind of terrorism is geared to facilitate foreign policy, covertly bringing pressure to bear on countries across the border through violence. The attack on the World Trade Center towers on 9/11 was a transnational terrorist incident because the victims were from ninety different countries, the mission had been planned and financed abroad, the terrorists were foreigners, and the sufferers were from various parts of the globe. India and many other countries across the world are victims of such terrorism. There are a number of countries with worrying rises in terrorism deaths. The country with the largest total increase in deaths from terrorism compared to the prior year is Afghanistan. Nine of the ten deadliest attacks in 2018 were in Afghanistan. In 2018, Afghanistan recorded the highest number of terror-related deaths (Fig. 3.2). Afghanistan has witnessed a substantial escalation in violence owing to a strengthened Taliban insurgency, increased presence of the Khorasan chapter of the Islamic state, and ongoing political instability.

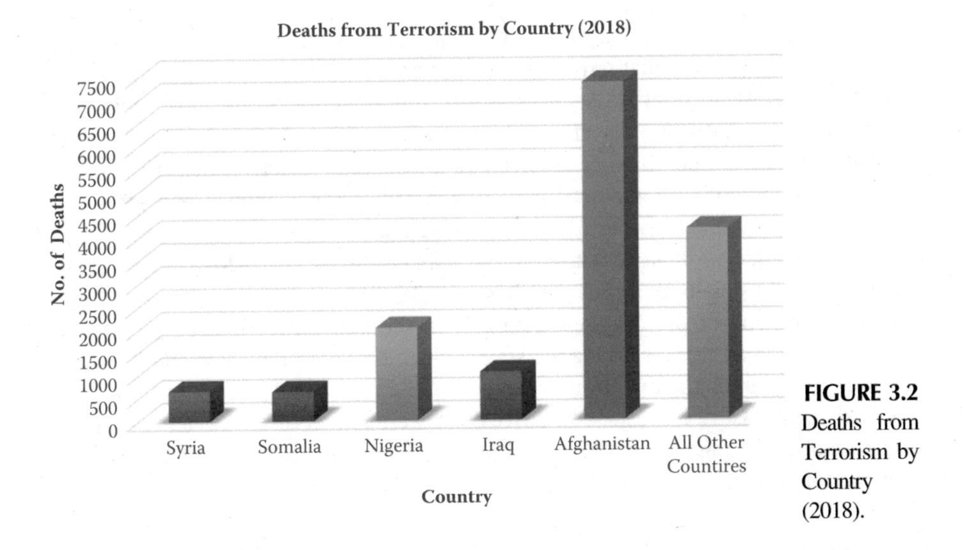

FIGURE 3.2 Deaths from Terrorism by Country (2018).

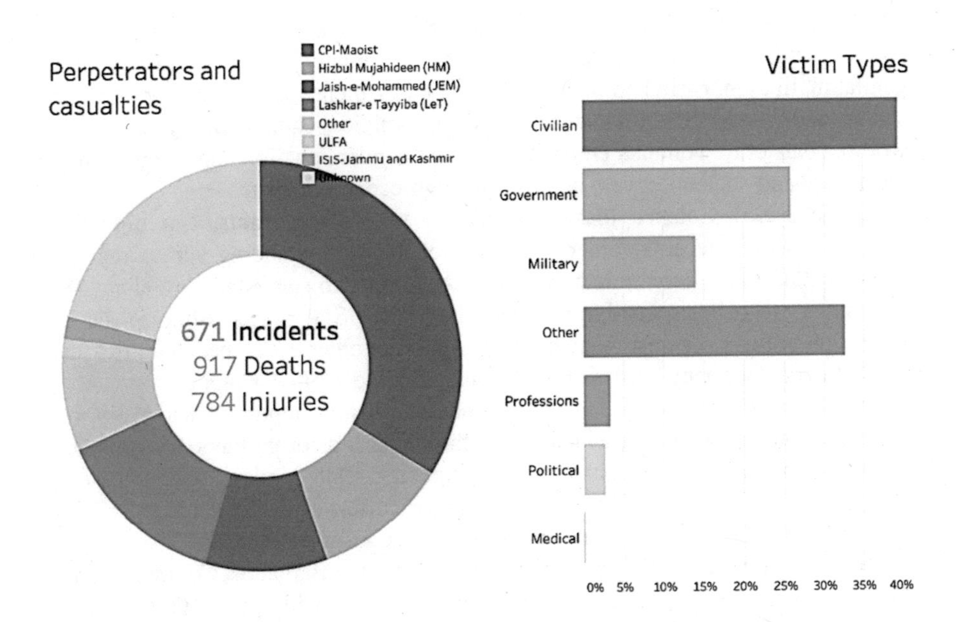

FIGURE 3.3 Terrorism Report India, 2018.

Post-9/11, India is highly vulnerable to terrorism by foreign terrorists. Terrorism Report India, 2018 reveals that there were 917 deaths and 784 injuries due to terrorist activities during 2018 (Fig. 3.3). There are number of training camps in Pak Occupied Kashmir (POK) in which terrorists are trained and sent inside Jammu and Kashmir and other parts of the country. Direct accessibility and the porous nature of the borders and a long coastline allow insurgents groups tactical flexibility. It is also worthwhile to mention that terrorists and subversive elements are using soft borders of India, touching countries such as Nepal and Bangladesh, to enter India. With Islamist and separatist groups all launching attacks in 2018, there is enhanced concern in the international community at terrorism emanating from Pakistan, including the continuing activities of internationally designated terrorist entities and individuals including Jamaat-ud Dawa (JuD), Lashkar-e-Taiba (LeT), Jaish-e-Mohammad, and Hizbul Mujahideen. India has suffered 8,473 deaths during 2001–2018 due to terrorist attacks. Terrorist activity in Pakistan has considerably reduced in recent years, but it still remains one of the deadliest countries in the world for violence by non-state groups. Khorasan chapter of Islamic state, Tehrik-i-Taliban Pakistan, and Lashkar-e-Jhangvi were among active groups that accounted for 15,908 terrorist-related deaths in Pakistan in last two decades. In 2016, there were a total of 13,488 terrorist attacks in the world, among which 4,573 occurred in Asia, which accounted for 24% of the international total (Hao et al., 2019).

3.2 CRIMES AGAINST WOMEN

Globally, one in three women experience either intimate partner violence or non-partner sexual violence during their lifetime. Intimate partner violence, female genital mutilation, early and forced marriage and violence as a weapon of war, and sexual and gender-based violence are major public health concerns across the world, a barrier to women's empowerment and gender equality, and a constraint on individual and societal development, with high economic costs.

Gender-based violence that threatens the well-being, rights, and dignity of women has only recently emerged as a global issue extending across regional, social, cultural, and economic boundaries. According to state statistics, about 18% of women are being sexually abused in the United States. According to the UN Report on violence against women, the condition in other developed countries such as Denmark, Germany, Spain, Switzerland, and the United Kingdom, etc., is no better. In the United States, the Department of Justice reported that every year hundred thousand women are battered by their husbands or partners. Even in some European countries, which ranks high in the gender-related index, reported cases of violence on women of domestic assault. Further, offenses of homicide recorded in some countries involved women killed by their spouses or lovers.

The data from developing countries such as Antigua, Barbados, Columbia, Chile, Ecuador, Guatemala, Sri Lanka, and others reveals widespread prevalence of physical and sexual abuse on women. A study on women from Japan carried out by Domestic Violence Group, reported physical, emotional, and sexual abuse. Studies from African countries Kenya, Uganda, and Tanzania also reveal that women are subjected to physical abuse at their homes.

The Universal Declaration of Human Rights and Convention on Elimination of all forms of Discrimination against Women (CEDAW) do enforce certain special rights and privileges for women. But it is saddening that only few countries have laws against domestic violence. Only a handful of countries have made marital rape a criminal offense and have passed laws on sexual harassment.

As per the estimate by UNWOMEN, 35% of women worldwide experienced either physical or sexual violence in 2017.

- A total of 2,554 women were victims of femicide in 2017.
- Women and girls together account for 71% of all human trafficking victims.
- Around 200 million women and girls alive today faced female genital mutilation in 30 countries.
- Twenty-three percent of female university students reported experiences of sexual assault or sexual misconduct.
- One in 10 women in the European Union report having experienced cyber-harassment since the age of 15.
- Forty to sixty percent of women report street-based harassment, mainly sexual comments, stalking/following, or staring/ogling.

Crime against women in India has shown various dimensions over thousands of years. They have also varied in intensity. Sometimes, crime against women is justified on the basis of culture, cuisine, caste, and color. However, such bases were more protuberant before independence. In India, nearly half of the population comprise women. Yet, they are dominated, suppressed, harassed, ill-treated, subjected to mental and physical violence, and sometimes even denied their basic human rights. They are the ones who are made to sacrifice and suffer without any right of complaining for it. The list of crimes that are committed against women vary from simple harassment, physical and mental torture, to even denying them the very right to exist. The National Family Health Survey (NFHS-4) suggests that 30% women in India in the age group of 15–49 have experienced physical violence since the age of 15. The report further reveals that 6% of women in the same age group have experienced sexual violence at least once in their lifetime. About 31% of married women have experienced physical, sexual, or emotional violence by their spouses. Crime against women has been a bane of India's development efforts. With arcane customs such as sex being a taboo in India, Sati, and Dowry, and the overall lower status of women further exacerbates these crimes. The crime rate in major states of India during 2018 has been classified and shown in Fig. 3.4. It reveals that the maximum crime rate against women is in Assam and Tripura, followed by West Bengal, Rajasthan, and Jammu and Kashmir. Nagaland and Meghalaya are among the top states where crime rates against women are at a minimum. Among the cities, Lucknow has the highest crime rate against women. Delhi, Indore, Jaipur, Kanpur, Ghaziabad, Nagpur, and Patna are other cities with high crime rates against women (Fig. 3.5). Rate of crime against women in different cities of India, in 2017, is presented in Fig. 3.6.

The tables presented are based on the projected incidence of crimes under various categories, assuming that these states maintain their torrid crime rate. A cursory look at the tables clearly indicates that Uttar Pradesh is leading in crimes against women in 4 of the 5 categories shown. Maharashtra and Andhra Pradesh are at number 2 and 3 in the total crimes against women and are significant contenders in other categories, too.

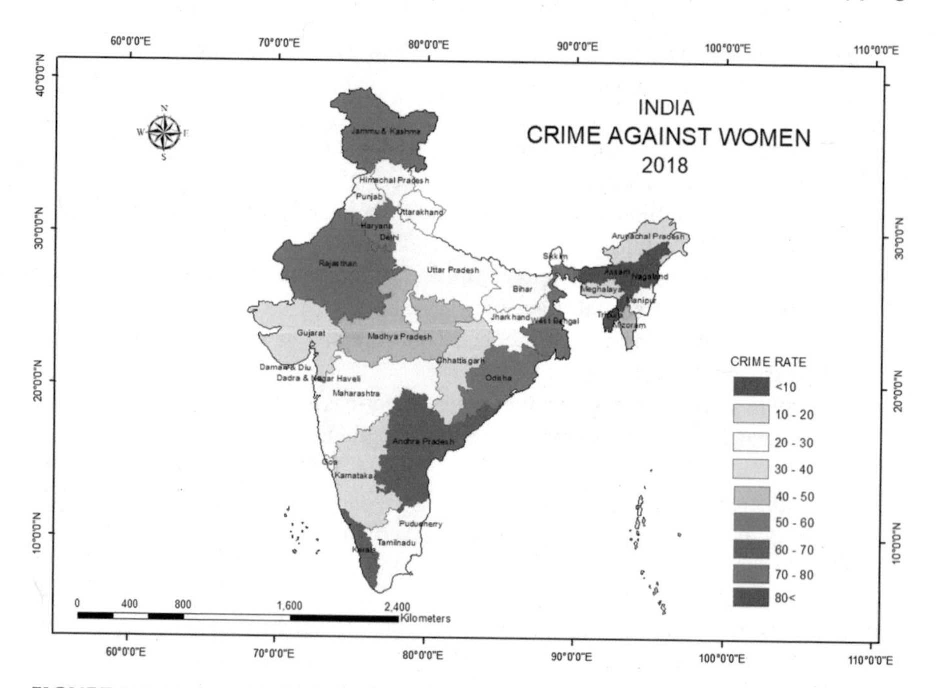

FIGURE 3.4 States with Highest Crime Rates Against Women in India, 2018.

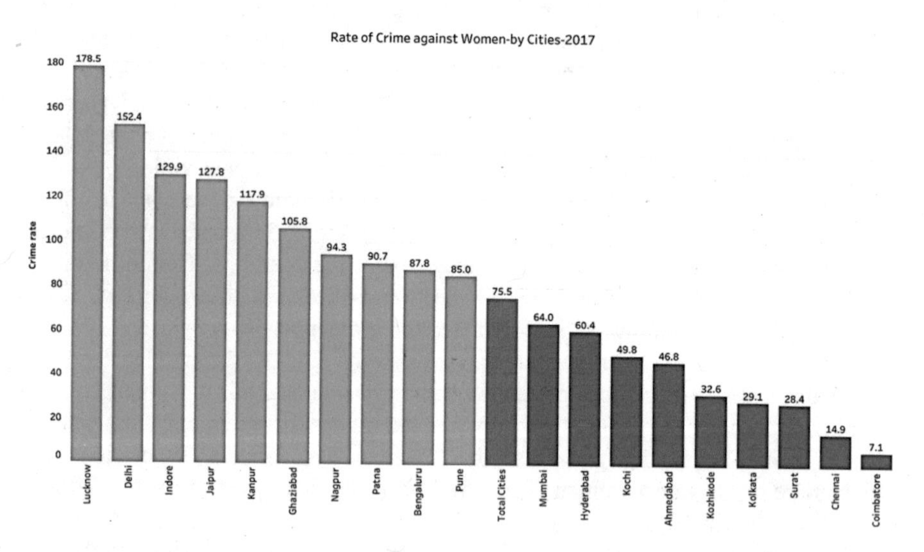

FIGURE 3.5 Rate of Crimes Against Women in Cities, 2017.

3.2.1 Projected Rapes in India

Rape in India has been described by Radha Kumar as one of India's most common crimes against women and by the UN's human-rights chief as a "national problem." Rape cases in India have doubled between 1990 and 2008. More than 32,500 cases of

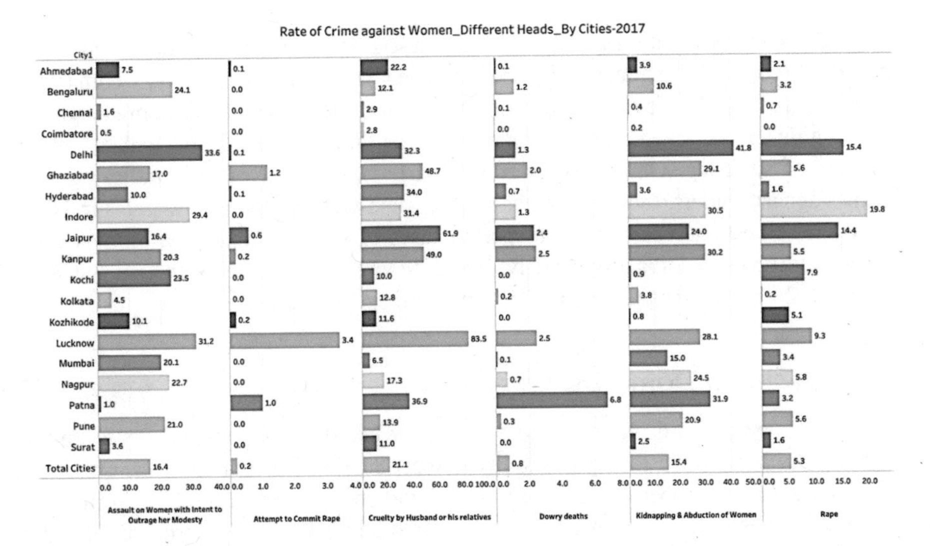

FIGURE 3.6 Rate of Crime Against Women in Different Cities of India, 2017.

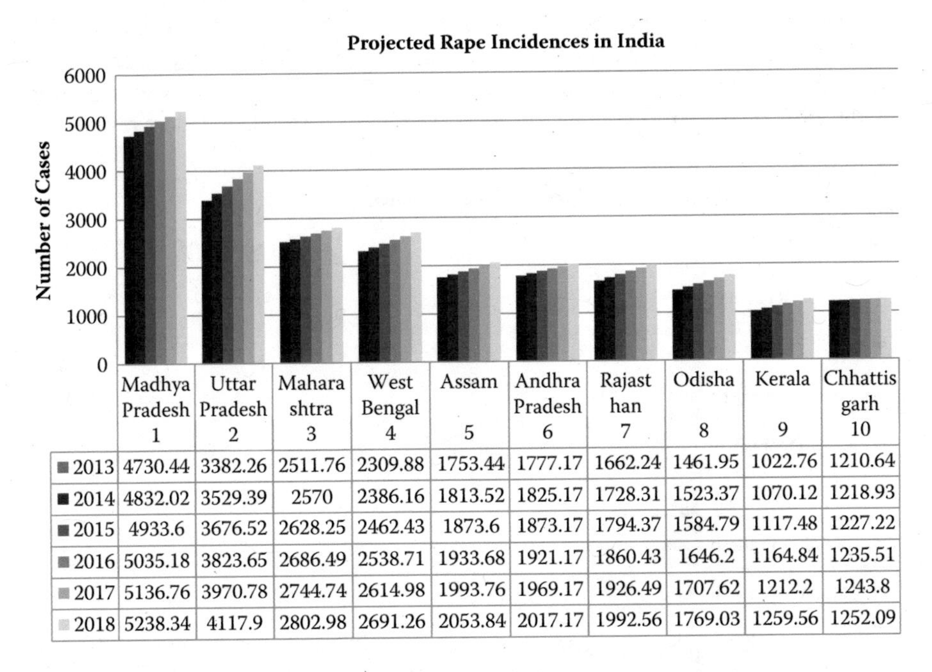

	Madhya Pradesh 1	Uttar Pradesh 2	Maharashtra 3	West Bengal 4	Assam 5	Andhra Pradesh 6	Rajasthan 7	Odisha 8	Kerala 9	Chhattisgarh 10
2013	4730.44	3382.26	2511.76	2309.88	1753.44	1777.17	1662.24	1461.95	1022.76	1210.64
2014	4832.02	3529.39	2570	2386.16	1813.52	1825.17	1728.31	1523.37	1070.12	1218.93
2015	4933.6	3676.52	2628.25	2462.43	1873.6	1873.17	1794.37	1584.79	1117.48	1227.22
2016	5035.18	3823.65	2686.49	2538.71	1933.68	1921.17	1860.43	1646.2	1164.84	1235.51
2017	5136.76	3970.78	2744.74	2614.98	1993.76	1969.17	1926.49	1707.62	1212.2	1243.8
2018	5238.34	4117.9	2802.98	2691.26	2053.84	2017.17	1992.56	1769.03	1259.56	1252.09

FIGURE 3.7 Projected Rape Incidences in India (Major States).
Source: NCRB.

rape were registered with the police in 2017, about 90 a day. Killing women after rape, and burning them alive are some of the heinous acts that are reported every day in India. According to NCRB statistics, Madhya Pradesh has the highest number of rape reports among Indian states, followed by Uttar Pradesh, Maharashtra, and West Bengal

(Fig. 3.7). Using a small sample survey, Human Rights Watch projects more than 7,200 minors (1.6 in 1,00,000 minors) are raped each year in India. Among these, victims who do report the assaults are alleged to suffer mistreatment and humiliation from the police. Minor girls are trafficked into prostitution in India; thus rape of minors conflates into a lifetime of suffering. Of the countries studied by Maplecroft on sex trafficking and crimes against minors, India was ranked the seventh-worst country. Few states in India have tried to estimate or survey unreported cases of sexual assault. The estimates for unreported rapes in India vary widely. The National Crime Records Bureau (NCRB) report of 2006 mentions that about 71% rape crimes go unreported.

3.2.2 Projected Dowry Deaths in India

Violence is perpetrated on an incoming bride by her spouse or marital family in retaliation for her and her family's inability to meet the dowry demands of the groom or his family. On a number of occasions, dowry violence may take the form of harassment that leads to death, known as "dowry death." In 1961, the government of India passed the Dowry Prohibition Act, making dowry demands in wedding arrangements illegal. Despite stringent law, a large number of cases of dowry-related domestic violence, suicides, and murders have been reported every year. Uttar Pradesh, Bihar, Madhya Pradesh, and Andhra Pradesh are among the top ten states where highest number of dowry-related deaths are registered (Fig. 3.8). Either the victims were burnt alive or forced to commit suicide over dowry demand. A report reveals that

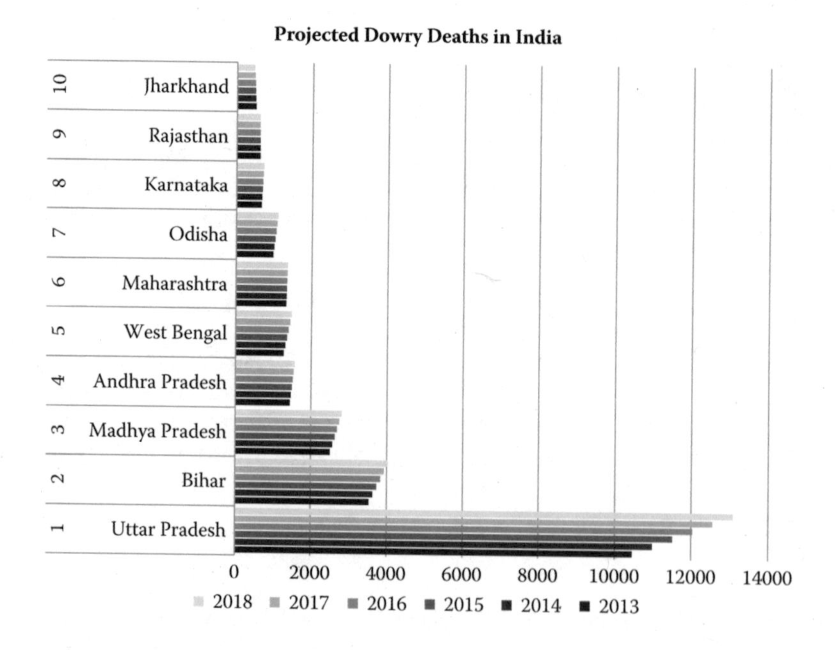

FIGURE 3.8 Projected Dowry Deaths in India (Major States).
Source: NCRB.

after registration of dowry deaths, police have charge sheeted around 93.7% of the accused, of which only 34.7% have been convicted.

3.2.3 Kidnapping and Abduction

Although India has passed a law against trafficking in 1956, the trafficking is a major social and economic issue. Girls are sometimes kidnapped from families and sometimes sold by parents or relatives because of economic issue. Poor public services in India adds to the social position of women. Most experts working in the field agree that when women and young girls are kidnapped, it is most often for purposes of sexual trafficking. Unsurprisingly, trafficking takes place largely in poverty-stricken states. Uttar Pradesh is among the top state in India where kidnapping and abduction is rampant and the state has maintained its top position in kidnapping and abduction since 2013 (Fig. 3.9). Bihar, West Bengal, Assam, and Andhra Pradesh are among the top five states where kidnapping and abduction goes unabated. There is hardly any development in rural areas of these states, which makes it easy for agents to lure desperate families into sending their girls away for work.

3.2.4 Cruelty by Husband and Relatives

The constitution provides equal opportunity to women in every sphere of life. She has equal opportunity for employment and protection at the workplace and the performance of her special biological functions such as childbirth, child rearing, etc. However, continuous abuse in honor of the house has been considered a part of culture to keep women in control. The most widespread form of cruelty against women is intimate partners' violence which has negative consequences on the

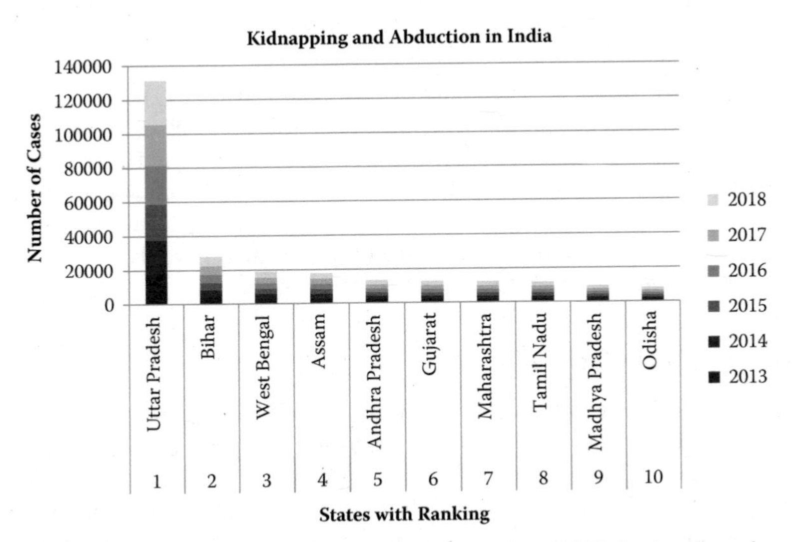

FIGURE 3.9 States with High Ranking in Kidnapping and Abduction Cases in India.

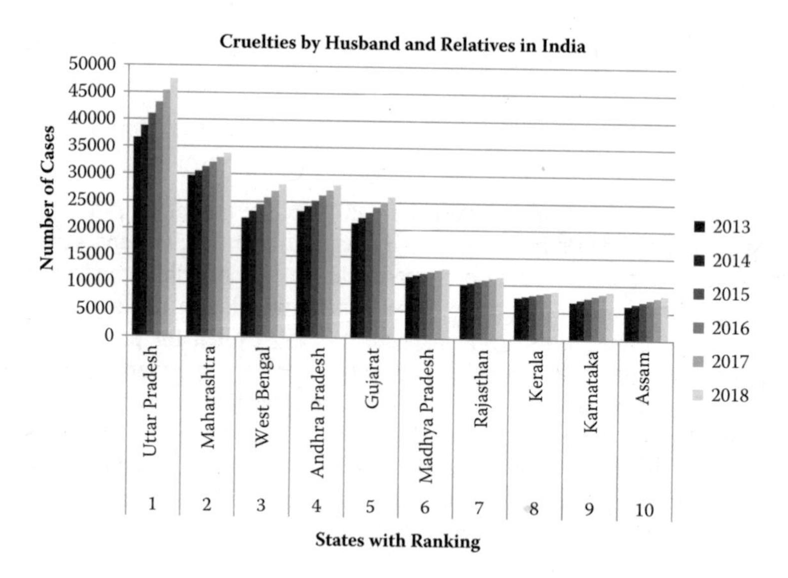

FIGURE 3.10 States with High Ranking in Cases of Cruelties by Husband and Relatives in India.

woman's health. Moreover, the psychological consequences are also huge, stemming from the paradox of the victim being abused by a member of the family with whom she expects to have a supportive, loving, and respectful relationship. Apart from this physical and mental torment, women are also denied economic independence. Uttar Pradesh again found itself among the top where the highest number of cases are reported against husbands and relatives for being cruel towards females of the house (Fig. 3.10). Maharashtra, West Bengal, Andhra Pradesh, and Gujarat are among the top five states where husbands and other family members show cruelty against women. According to the latest data by the National Commission For Women, March 2020, a staggering 257 cases of crime against women for a duration of 10 days (23 March to 1 April) of lockdown was recorded as compared to 116 cases that was recorded for a duration of 7 normal days (2 March to 8 March). Assumptions were that the 21-day-long lockdown may keep women safe from coronavirus, but it has pushed a lot of them towards something equally grievous—domestic violence and sexual harassment.

3.3 JUVENILE DELINQUENCY

The word "juvenile" originates in a Latin word *juvenis* that means young. A juvenile or child means a person who has not completed 18 years of age. Children are our greatest national resource. They become the leaders of the country, the creators of national wealth, and who care for and protect the human community of the land to which they are rooted. These children across the world develop at different rates and develop different world views. They increase their ability to think abstractly and develop their own views regarding social and political issues. They develop the ability to indulge in long-term planning and goal setting. There is also a tendency of making comparisons of self with others. They yearn for a separate identity and independence from parents. This is the age when peer influence and acceptance becomes very important. They also develop strong romantic/sexual ideas, and tend to show indulgence in love and long-term relationships. However, these are normal changes and there are no anomalies generally. Problems arise when these juveniles develop delinquent tendencies, and get into law and order problems. There appears to be a very strong relationship of crime/deviance with age; The general observation is that criminality/delinquency peaks in adolescence and diminishes with age; this pattern of crime is common across historical, geographical, and cultural contexts. Indulgence in conventional crimes is more widespread in teenagers and young adults. Most of these offenders disengage from crime after a brief career in crime. Juvenile crimes have become such common phenomena that they raise serious concerns in any nation; for example, the share of crimes committed by juveniles to total crimes reported in India has increased particularly in recent years. Juvenile crime, formally known as juvenile delinquency, is a term that defines the participation of a minor in an illegal act. And juvenile justice is the legal system that aspires to protect all children, bringing within its ambit the children in need of protection, "besides those in conflict with law."

In common terminology, a juvenile is a child who has not attained a certain age at which he/she can think rationally and often understand the consequences of his/her act. Hence, the juvenile can't be held liable for his/her criminal acts. A juvenile delinquent may be regarded as a child who has allegedly committed/violated some law, under which his/her act of commission or omission becomes an offense. Under the Indian Laws, Section 2 (k) of the Juvenile Justice (Care and Protection of Children) Act, 2015 (referred generally as JJ Act), a juvenile is a person who is below 16 years. Prior to JJ Act of 2015, the age bar for juveniles was 18 years (Juvenile Justice (Care and Protection of Children) Act, 2000, 2006, 2012). In fact, the age of the juvenile under the Indian legislations has taken variation in temporal and spatial perspectives. It varies from 14 to 18 years under different laws and different Indian states.

3.3.1 International Scenario

Juvenile delinquency has become a cause of social concern all over the world. It is a universal phenomenon, in the sense that it is found in all human groups. Perhaps there is no society who has a pause as far as misbehavior or crime by individuals, more so among the younger ones that is children. The nature of human behavior is such that violation of standards of behavior or actions are sometimes hardly kept intact, in tune with that of norms, regulations, code of conduct, values, etc., of the society. The problem of human misbehavior has been observed and felt throughout the human world and at all levels of strata. However, the causation of delinquency, nature and extent of juvenile delinquency, genesis of delinquency, forms of delinquency, and patterns of delinquency vary from time to time, place to place, and well within a community, society, region, and nation. Shoemaker (2009) provides an overview of the history of the problem. According to him, juvenile delinquency is purported to have begun in Europe in the 7th century. Puritanism was one of the major ideologies of the time. This ideology advocated strict and acceptable individual and collective behavior, especially in matters related to religion. As early as the 17th century, children were treated as adults and, likewise, were held accountable for their actions. Children are said to have been deliberately exposed to sexual activity, hardships, and adult behaviors with the express purpose of initiating them into adulthood. They were also harshly punished for any misdemeanors because it was believed that corporal punishment was good for discipline.

In 1704, Pope Clement XI first introduced the idea of the "instruction of profligate youth in institutional treatment". Then, Elizabeth Fry established a separate institution for juvenile offenders. Subsequently, in Britain, the Reformatory Schools Act and Industrial Schools Act were brought a statute book. The first Juvenile Court was established in 1899 in Chicago under the Juvenile Offenders Act and the first Juvenile Court in England was set up in 1905. First probation law was enacted in the state of Massachusetts, USA, in 1878 and in England in 1887. The second and sixth UN Congress on Prevention of Crime and Treatment of Offenders in 1960 and 1980 discussed in detail the problem of juvenile delinquency. They decided that there should be the standard minimum rules to address the problem. The issue was again discussed in Beijing in 1985, which examined the Standard Minimum Rules for the Administration of Juvenile Justice. In 1989, the UN Convention on Rights of the Child (CRC) draw attention to four sets of civil, political, social, economic and cultural rights of every child. The convention provides the legal basis for initiating action to ensure the rights of children in society. A report by the Geneva Convention for the Rights of The Child (2007) attests to the view that juvenile misconduct is a global phenomenon. A report on juvenile delinquency drafted for the European Parliament expresses the same sentiment. The report was adopted by the European Parliament in 2007. This challenge begs the question: What is being done to address juvenile delinquency at its onset before it escalates into serious crime and possible persistent misbehavior? Marte (2008) also concurs with other researchers that early childhood delinquency has the likelihood of persisting throughout adolescence and adulthood. He urges further that the etiology of antisocial behavior is crucial to understanding problem behavior so that appropriate interventions can be developed.

3.3.2 Indian Scenario

India has a long history of providing separate treatment for juvenile offenders. Differential treatment for children can be traced as far back as the Code of Hammurabi in 1790 BC, the responsibility for their supervision and maintenance being vested on the family. During the colonial regime, in 1843, the first center for those children, called "Ragged School," was established by Lord Cornwallis. The period between 1850 and 1919 was marked by social and industrial upheavals. The Apprentices Act, 1850, 2, came into force, which required that children between the ages of 10–18 convicted in courts, to be provided vocational training as part of their rehabilitation process. For the treatment of juvenile delinquents, the next landmark legislation was the Reformatory School Act, 3 1876 and 1897. Under the Act, the court could detain delinquents in a reformatory school for a period of 2–7 years but after they had attained the age of 18 years, the court could not keep them in such institutions. The Act of Criminal Procedure, 1898, provided special treatment for juvenile offenders. The code provided probation for good conduct to offenders up to the age of 21. Thereafter, the Indian Children Act (Indian Jail Committee 1919–1920) came into existence, where individual provincial governments chose to enact separate legislation for juvenile in their respective jurisdictions; provinces of Madras, Bengal, and Bombay passed their own children acts in 1920, 1922, and 1924, respectively. These laws contained provisions for the establishment of a specialized mechanism for the treatment of juveniles.

In post-independence period, the juvenile justice policy in India is structured around the constitutional mandate prescribed in the language of Articles 15 (3), 21, 24, 39 (e) and (f), 45, and 47. International convention, such as the UN Convention on the Rights of the Child (CRC) and the UN Standard Minimum Rules for Administration of Juvenile Justice (Beijing Rules), are also considered. The Juvenile Justice Bill (JJB) was first introduced in the Lok Sabha in 1986, and the Central Children Act (CCA) was replaced by the Juvenile Justice Act. The Juvenile Justice (Care and Protection of Children) Act, 2000, brought in compliance of Child Rights Convention 1989, repealed the earlier Juvenile Justice Act of 1986 when India signed and ratified the Child Rights Convention 1989 in 1992. This Act has been further amended in the years 2006 and 2010. It deals separately with two categories of children. Juveniles accused of a crime or detained for a crime are brought before the JJB, and not in a regular criminal court. The JJB consists of a first-class magistrate and two social workers, at least one of whom should be a woman. A Special Juvenile Police Unit (SJPU) shall be set up in every police station. A child is usually brought before the JJB by a police officer or person from the SJPU. The police have 24 hours to produce a child before the court once he/she is arrested. Once the child has been brought before the JJB, the child is registered into the closest observation home. A probation officer will be deputed as the juvenile's guidance for the accused. JJB are meant to resolve cases within a 4-month period. Maximum sentence for a juvenile who has broken the law is 3 years in a protective home, no matter how serious the crime is.

The roles of after-care organization is very important. These organizations are for the care, guidance, and protection of juveniles in conflict with law or children in need of care and protection who have completed their terms in the special homes. After-care organization shall enable such children to adapt to the society and encourage them to live a normal life. Based on the resolution passed in the conference of Chief Justice of India 2009, several high courts constituted "Juvenile justice Committees" to be headed by sitting judges of high courts. In India, juvenile delinquency is increasing at an alarming rate. The involvement of the juveniles in serious offenses such as murder, attempt to murder, kidnapping, and abduction has raised concerns in the nation. After the December 2012 gang rape in Delhi, many debates and discussions pointed to the softer approach of the Juvenile Justice System to serious offenses. It has been found that the youngsters can be as brutal as the adults, which forced the people to reanalyze the definition and approach to juvenile delinquents in India. Due to access to the Internet, aspirations of adolescents and adults are becoming at par (Ghosh, 2013). The National Crime Records Bureau (NCRB) data indicates that there has been an increase in crimes committed by juveniles, especially by those in the 16–18 years' age group. There is an increase in number of cases registered against juveniles in conflict with law (Agarwal, 2018). Agarwal found out that from 2005 to 2015, this number has increased from 18,939 to 31,396 (in category of Against Juveniles in conflict with law) and from 1,82,2602 to 2,94,9499 (under total cognizable IPC Crimes). Figures are pretty alarming as far as crime by juveniles under IPC and SLL Crimes by Age Groups and Sex during 2015 under different age categories is concerned. It is evident that involvement of juveniles in the age group of 16–18 years is very high. Various reasons can be cited for this rise in juvenile delinquency. According to the psychiatrists and women' rights activists, easy access to pornography and changing food habits can be attributed as a cause for this change in behavior of juveniles, who show rising involvement in sexual offenses (Mishra, 2013). Youngsters are not able to control their biological impulses prompted by hormonal changes. The family as a basic unit of human society is getting weaker, particularly in urban areas, with lesser family control on children. The community networking and involvement in affairs of individual members has also become slack. Peer groups are becoming less active, with the youth mainly spending time indoors watching television, or playing games on mobile or computer devices. The mind of the youth is a place where a lot of information keeps gathering which may have no substance. There are fewer options of venting out their frustrations and negativities, leading to pent-up teenage aggression.

3.3.3 Types of Delinquency

Delinquency exhibits a variety of styles of conduct or forms of behavior. Each of the patterns has its own social context, the causes that are alleged to bring it about, and the forms of prevention or treatment most often suggested as appropriate for the pattern in question. Broadly speaking, there are four types of delinquencies:

- Individual delinquency
- Group-supported delinquency
- Organized delinquency
- Situational delinquency

3.3.3.1 Individual Delinquency

This refers to delinquency in which only one individual is involved in committing a delinquent act and its cause is located within the individual delinquent. Most of the explanations of this type of delinquent behavior come from psychiatrists (Venkatachalam and Aravindan, 2014). Their argument is that delinquency is caused by psychological problems that primarily stem from defective/faulty/pathological family interaction patterns. Psychologists have identified several factors for children delinquency. Children's behavior is the result of genetic, social, and environmental factors. Some of the individual factors identified as risk factors in juvenile delinquency are early antisocial behavior, emotional factors, and cognitive development and hyperactivity. These factors are frequently interrelated, yet the underlying mechanism of how this occurs is not fully understood. Antisocial behavior at an early age may be the best predictor of later delinquency. Another factor that affects a person's character and is considered by criminologists is hereditary factors. As humans are financially the heir of relatives, they may also inherit their talents and good and bad attributes of their ancestors, which are transferred to them through inheritance. Gender is another important risk factor concerning delinquent or antisocial behavior, especially in terms of aggressive behavior. Emotional factors may also at a later stage contribute to delinquent behavior.

3.3.3.2 Group-Supported Delinquency

This is a form of juvenile delinquency behavior where criminal activities are carried out by a group of children rather than a single child. The cause of this form of juvenile delinquency is neither children's family issues nor personality of the child but in the structure of the immediate neighborhood society with emphasis either on the ecological areas where delinquency prevails. Thus, group-supported juvenile delinquency comes about due to association of non-delinquent children with their already delinquent peer friends. At least one delinquent that commits crime in a group is usually unwilling to participate in the act. It is not in the personality of the delinquent to commit a crime but goes along with friends or peer pressure. The culture of delinquency may be in the culture of the neighborhood that the juvenile is staying in or at his home. The study of Shaw and McKay (1931) talk of this type of delinquency.

3.3.3.3 Organized Delinquency

This type refers to delinquencies that are committed by formally organized groups. These delinquencies were analyzed in the United States in the 1950s and the concept of "delinquent sub-culture" was developed. This concept refers to the set of values and norms that guide the behavior of group members to encourage the commission of delinquencies, award status on the basis of such acts, and specify typical relationships to persons who fail outside the groupings governed by group norms. The group gains its status from committing criminal activities. For instance, take the example of drug trafficking that is done by the youngsters in Mumbai and other cities of India. This is common among the street kids whom the organized groups pay in drugs to deliver their packet of drugs in various parts of the city.

3.3.3.4 Situational Delinquency

The above-mentioned three types of delinquencies have one thing in common. In all of them, delinquency is viewed as having deep roots, whereas situational delinquency provides a different perspective. Matza (1964) is one scholar who refers to this type of delinquency. Here, the assumption is that delinquency is not deeply rooted, and motives for delinquency and means for controlling it are often relatively simple. A young man indulges in a delinquent act without having a deep commitment to delinquency because of less developed impulse control and/or because of weaker reinforcement of family restraints, and because he has relatively little to lose even if caught. What impels these kids to break the law and commit beastly acts is difficult to understand, hence the concept of situational delinquency is undeveloped and is not given much relevance in the problem of delinquency causation.

3.3.4 Reasons for Juvenile Crimes

Interdisciplinary studies on juvenile delinquency reveal that across the world, many behavioral changes occur in the juveniles, which are related to the sudden changes in their body due to hormonal surge, associated with puberty. The changes are most apparent in physical parameters, such as change in height and weight of the adolescents, and are soon followed by other sexual and physical changes of maturity. These physical changes are accompanied by mental changes also.

3.3.4.1 Biological Factors

The biological explanations suggest that individuals are influenced by their biological/ genetic makeup. Biological theories attempt to explain crime committed by juveniles primarily in terms of factors within the criminal or delinquent. Accordingly, juveniles are not exactly the captives of biological designing, but it does render these individuals inclined towards delinquent tendencies. Investigations made by social scientists indicate that the roots of criminal behavior is evident in bodily characteristics such as slanting forehead, large jaws, heavy brows, etc. Similarly, hormonal changes in the body of the juveniles are responsible for their impulsive and rebellious behavior. Researchers takes into account other factors such as climate, race, alcoholism, education, wealth, etc., in their search for the cause of crime and delinquency. Another biological factor associating physical characteristics with criminal behavior is the chromosome. Biological problems such as speech and hearing problems, irritation, excessive strength, etc., may also lead to delinquency. Ecological/environmental and economic parameters together also play important trigger points in lives of the juveniles.

3.3.4.2 Socioeconomic Factors

Social norms are certain behavioral techniques that are formed according to social values of society. Social norms are certain behavioral techniques that are formed according to social values of society and it is by their observance that the society follow certain rules. Official regulations, legislation, jurisprudence and religious customs, ethnical practices, and the like are considered as the norms of society. However, sometimes, the juveniles develop a delinquent sub-culture due to cultural deprivation and status frustration that they go through. Thus, if certain morals are upheld by society, juvenile delinquents give up these values and try to excel in the areas of toughness, over-smarting the others and indulge in things that give them excitement. Mobility, cultural conflicts, and family background are important drivers of juvenile delinquency among youth. Juvenile delinquency is also driven by the negative consequences of social and economic development, in particular economic crises, political instability, and the weakening of major institutions. Socio-economic instability is often linked to persistent unemployment and low incomes among the young, which can increase the likelihood of their involvement in criminal activities.

3.3.4.3 Psychological Factors

There can be a strong link between psychological condition of the youth and delinquent tendencies. There are psychological explanations to delinquency, which can be well understood through Freudian concepts of id, ego, and superego. When the id (the instinctive element of individual's personality) becomes too strong, and the superego becomes weak (the socially taught element of personality), the ego develops into an antisocial person. Sometimes when the self-control and social control through primary groups becomes weak, the juveniles develop delinquent tendencies. The breakdown of the social institutions has also been connected to deviance and delinquency (Knoester and Haynie, 2005).

3.3.4.4 Socio-Environmental Factors

Mobility

Mobility is responsible for crime causation in the society. The rapid growth of industrialization and urbanization has led to expansion of means to communication, travel facilities and propagations of views through press and platform (Choudhary, 2017). The ongoing process of urbanization particularly in developing countries is contributing to juvenile involvement in criminal behavior. The basic features of the urban environment foster the development of new forms of social behavior deriving mainly from the weakening of primary social relations and control, increasing reliance on the media at the expense of informal communication, and the tendency towards anonymity. Many geographical analysis suggests that countries with more urbanized populations have higher registered crime rates than do those with strong rural lifestyles and communities. This may be attributable to the differences in social control and social cohesion. Rural groupings rely mainly on family and community control as a means of dealing with antisocial behavior and exhibit lower crime rates, whereas urban-industrial societies tend to resort to formal legal and judicial measures, an impersonal approach that appears to be linked to higher crime rates. Migration of persons to new places where they are strangers offers them opportunity for crime as chances of detection are minimized considerably. This type of pattern is generated by the high population density, degree of heterogeneity, and number of people found in urban contexts.

Cultural Conflicts

In a dynamic society, social change is an inevitable phenomenon. The impact of modernization, urbanization, and industrialization in a rapidly changing society may sometimes result in social disorganization and this may lead to culture conflicts between different valves of different sections of society (Choudhary, 2017). Crime in many instances is a product of culture conflict between the values and norms of a certain sub-culture in a given society and those of the general culture. Cultural conflicts can be in two forms: primary and secondary. The primary cultural conflict results when the norms and value system of different cultures clash with each other, whereas secondary cultural conflict grows out of the process of social differentiation, which characterizes the evolution of one's own culture. Immigration also affects the crime rate of a place. Immigrants arriving from other places who bring their home culture with them do not necessarily fit in with those already established and, in some cases, cause conflict.

Environmental Factors in the Family

Family is the first agent of socialization. It is responsible for the basic love, care, sense of security, physical, mental, moral, and emotional development of a child. Thus the family environment is overwhelmingly positively correlated with juvenile delinquency, both as a precursor of, and buffer against, youth misconduct. Parental warmth, supervision, support, and involvement help children to overcome challenges thrown upon them by society. A negative family environment also produces behavioral, emotional, and affective problems in juveniles. In families where parents are cruel or violent, gamble, or do antisocial activities, children emulate these behaviors and use them in other contexts.

Structural Breaks in the Family

Structural breaks in the family, except in case of death, are always seen to be preceded by daily parental quarrels. Much tension and disruption of peaceful living have quite a traumatic effect on the children. When parental relationships deteriorate into regular fights, break-ups, and step-parenting, a child is deprived of his basic needs to be loved, recognized, and understood. Failure to meet these basic needs often results in dissatisfaction, and delinquency is likely to take root. The development of a child's personality is, thus, greatly molded by what he learns during his early life. Healy and Healy (1917), Burtheal (1925), Leeper (1925) have pointed out that this situation may be considered a very important issue in delinquency. Healy reports 45 to 52%, Burt 59%, and Leeper 79% of cases of delinquency where "home disrupted during the childhood of the individual" occurs in the case histories. British and American investigations disclose that nearly 50% of the delinquents come from broken homes.

Substance Abuse

All of the circumstances mentioned are responsible for substance abuse among the participants as perceived by them. The participants admitted that they took over substance use as a coping mechanism to deal with multiple stressors resulting from adverse family and living conditions (Mhavan, 2017). Failure in schooling also instigated them to consume various substances which started with experimenting as a part of curiosity and ended up in addiction. Deviant peer association was a measure factor that debilitated the sense of judgment of the participants and got heaved into substance use. This finding also supports the central thrust of general theory of crime by Gottfredson and Hirschi (1990), which is self-control. If the participants had self-control, it would help them to avert from their deviant peers and resultantly from substance abuse. The participants also revealed that they gained courage from substance use and were able to execute the delinquency act firmly.

Peer Association

Friendships with corrupted and profligate individuals is among the other factors which affect the delinquency of children (Farhadian, 2016). Adolescents often imitate their friends' behavior and are strongly influenced by them. An adolescent, who is rejected by their family, will turn to friends of the same age to compensate emotional and psychological deficiencies and gain support and appreciation. They are looking for people who are like them. It is therefore likely to get involved in anti-social and misdemeanor acts by the encouragement and influence of corrupted friends. Also, in order to get popularity and position among the friends, the individual tries to be like the others.

Mass Media

Despite all the benefits of the development of the mass media, it has some disadvantages as well. With the development of television, magazines, cinemas, computers, and the Internet, such devices have turned into criminal training schools for children and adolescents, who have a very high talent for learning and making imitations. Display of combat and war movies, increased scenes of proficient robbery and obscene scenes, each in its own way have caused a negative impact on them and their orientation towards the aforementioned criminal acts. Obviously, familiarity with these issues, day by day increases the rate and extent of juvenile crime (Putwain and Sammons, 2002).

Neighborhood

Neighborhood influences also have much to do with the nature of crimes in the particular locality. Thickly populated areas offer frequent opportunities for sex offenses and crimes relating to theft, burglary, kidnapping, cheating, deceit, etc. (Choudhary, 2017). The cases of pickpocketing are common at railway stations and bus stops and other halt stations. Vehicle theft by youngsters is too common. Another significant feature of delinquency is certain antisocial activities in the neighborhood. These include prostitution houses, gambling houses, brothels, and similar other dubious character institutions. The cinema theaters, swimming pools, sports grounds, and race courses generally offer a favorable atmosphere for delinquents. Studies across various time frames have stated that there exists a relationship between neighborhood and delinquency. Slum neighborhoods provide a good ground for the growth of juvenile delinquency.

3.4 CRIME IN NCR

Freedom from violence, as an aspect of the quality of life, is a neglected issue in development studies. Most people would rather avoid being mugged, beaten, wounded, or tortured, and it is also nice to live without fear of these traumatic experiences. Violence also affects human well-being in indirect ways, as when armed conflicts undermine economic growth or the functioning of public services. If development is concerned with improving the quality of life, the issue of violence should be a major interest of the discipline. Yet, it tends to receive little attention outside specialized circles. Protection from violence, however, is not a convenient by-product of economic growth and indeed there are spectacular cases of violence rising against a background of rapid improvement in per capita income and other development indicators. Dealing with violence in a society is, therefore, intrinsically a matter of public action. So it is essential to have a careful investigation of the causes of violence. Crime in India, an annual publication of the government of India (Ministry of Home Affairs), presents district-level data on a range of "crimes" such as murder, rape, kidnapping, theft, burglary, and arson (Dreze and Reetika, 2000).

India has experienced several noteworthy demographic changes over the last decade. The 2011 Census also indicated that a large part of the urban population continued to be concentrated in Class-I (1 lakh and above) Urban Agglomerations (UA)/Towns (70%), of which around 43% were in million-plus UAs/cities alone. Interestingly, while the growth rates of some of its mega cities (10 million plus population), especially Delhi, Mumbai, and Kolkata, reduced drastically in the last decade, the number of million-plus cities increased from 35 as per the 2001 Census to 53 as per the 2011 Census, with 18 new UAs crossing the million mark, implying the rapid growth of other large urban centers (Cities Alliance, 2015).

3.4.1 NCR Constituent Areas

National Capital Region (NCR) is a unique example of inter-state regional planning and development for a region with NCT-Delhi as its core (Fig. 3.11). The NCR as notified covers the whole of NCT-Delhi and certain districts of Haryana, Uttar Pradesh, and Rajasthan, an area of about 55,098 square kilometers, covering a population of about 581.5 lakhs (Table 3.1).

Keeping in view the exponential growth of Delhi and surrounding area, it was considered that the planned growth of Delhi is possible only in a regional context. In fact, the need for regional approach was felt as early as 1959 when the draft Master Plan for Delhi was prepared. Thereafter, the Master Plan of 1962 recommended that a statutory National Capital Region Planning Board should be set up for ensuring balanced and harmonized development of the region. The setting up of the statutory board in 1985 and coming into operation of the first statutory Regional Plan-2001, were the important achievements in the balanced development of the National Capital Region (NCRPB, 1999).

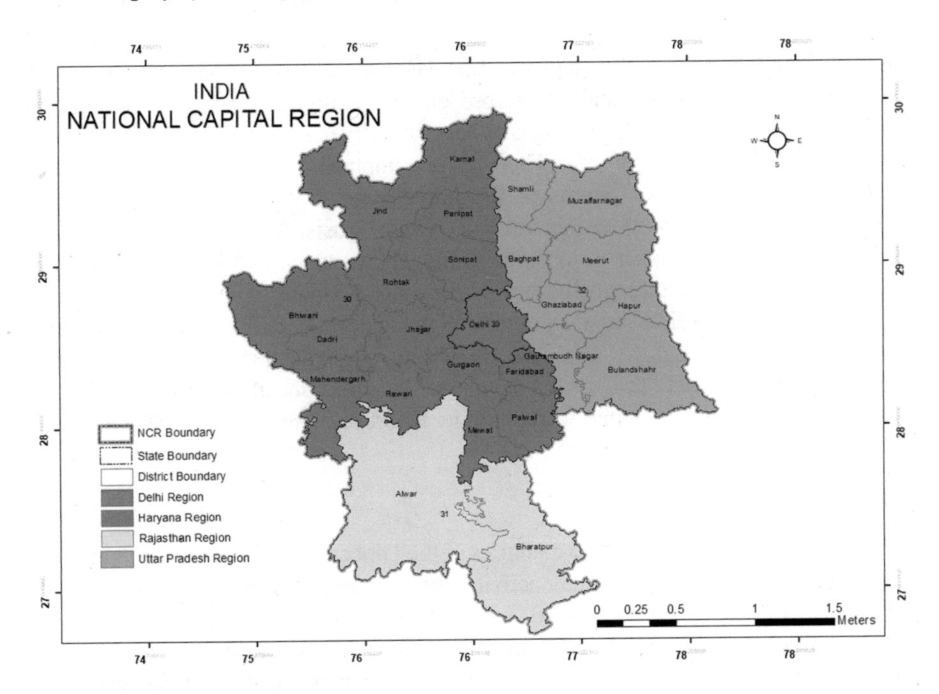

FIGURE 3.11 National Capital Region (NCR), India.
Source: NCRPB, 2018.

TABLE 3.1
Sub-region Wise Area Details of NCR

Sub-Region	Name of the Districts	Area (in sq. kms.)
Haryana	Faridabad, Gurgaon, Mewat, Rohtak, Sonepat, Rewari, Jhajjhar, Panipat, Palwal, Bhiwani (including Charkhi Dadri), Mahendragarh, Jind, and Karnal (**Thirteen Districts**).	25,327
Uttar Pradesh	Meerut, Ghaziabad, Gautam Budh Nagar, Bulandshahr, Baghpat, Hapur, Shamli, and Muzaffarnagar (**Eight Districts**).	14,841
Rajasthan	Alwar and Bharatpur (**Two Districts**).	13,447
Delhi	Whole of NCT Delhi.	1,483

Source: NCRPB, 2018.

3.4.2 Crime Scenario at National Level

The inability of the system to deliver justice and maintain the rule of law has led to an uptick in violence, which according to the Institute for Economics and Peace, has cost India an equivalent of 9% of its GDP. The rule of law is also irrevocably linked with raising the individual's quality of life. The Ease of Living Index (2018), which assesses ease of living standards across cities, highlights the importance of improving governance, infrastructure and service delivery, all of which have a direct bearing on the quality of life. There is an emphasis on "safety and security," quantitatively assessed in terms of the prevalence of violent crime, particularly against vulnerable groups and surveillance (Tata Trusts, 2019).

Crimes are much higher in mega cities (having population of 10 lakh/1 million or more) compared to either small cities or rural areas. Some of the mega cities have comparatively high crime rate than others. High incidents of crimes in mega cities may be due to various factors such as high density of population, greater information availability/flow, greater degree of anonymity in big cities, social milieu of urban slums, etc. NCRB analyzed the crime scenario in its report in 53 mega cities (NCRB, 2015).

3.4.3 Crime Scenario in NCR

India was ranked 136 out of 163 countries in the Global Peace Index in the year 2018. However, risks posed by criminal activity ranks seventh in India Risk Survey 2018, compared to India Risk Survey 2017 ranking at the eighth position. Delhi, Kochi, Indore, Bhopal, and Gwalior are among the cities with the highest crime rate in the country, with Delhi having a crime rate of 182.1 compared to the national average of 77.2. However, increasing awareness among citizens and the incorporation of people-friendly mechanisms for crime reporting, there has been a marked increase in the number of crimes being reported. The government and police have made use of social media and digital platforms to reach more people and ensure more efficient policing. India Risk Survey 2018 finds that threats to security and safety posed by violent crimes, such as kidnapping and murder, are the biggest risk to people. In India, in 2016, IPC crimes have increased by 0.9% and Special and Local Laws (SLL) crimes have increased by 5.4% over 2015. Delhi registered the highest crime rate with 974.9 per 1 lakh population, followed by Kerala with 727.6 and Madhya Pradesh with 337.9. The risk of crime is a serious concern in India, affecting perceptions and hampering global investments and relations (FICCI, 2018).

NCR is India's largest and the world's fourth-largest agglomeration. It is home to major companies in the IT, offshoring/outsourcing, IT enabled services (ITES), manufacturing, and service sectors. Rapid urbanization has led to certain factors which have a bearing on the crime rate in the city of Delhi. The crime in Delhi is increasing every year (Fig. 3.12). The large expansion of new colonies such as Dwarka, Rohini, etc., and mushrooming of numerous unplanned colonies are important criminogenic factors which have a direct effect on crime, particularly the street crimes such as robberies and snatchings (Upadhyaya, 2017).

Crime is a manifestation of myriad complex factors. The genesis of crime can be traced to interplay of various social, economic, demographic, local, and institutional factors. Crime increases with population. More population as compared to the previous year will result in increased crime. A significant increase in cognizable crime rate may be traced in NCR (Fig. 3.13). Crime per lakh population (crime rate) may therefore be a better indicator (NCRB, 2018).

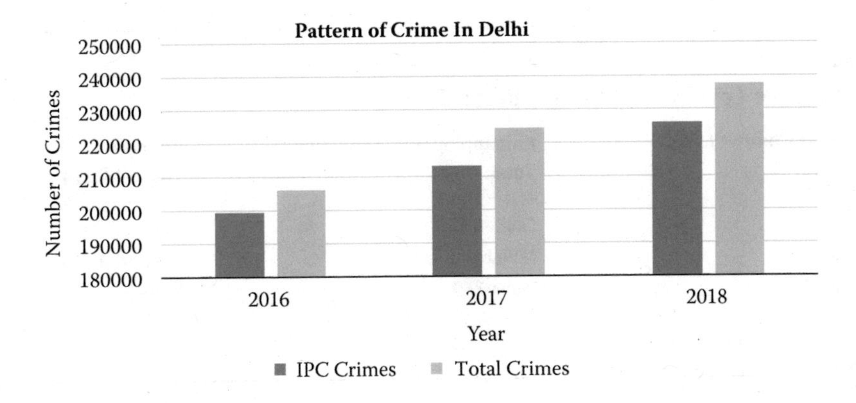

FIGURE 3.12 Graph Indicating the Pattern of Crime in Delhi, India.
Source: Crime in India, NCRB (2018).

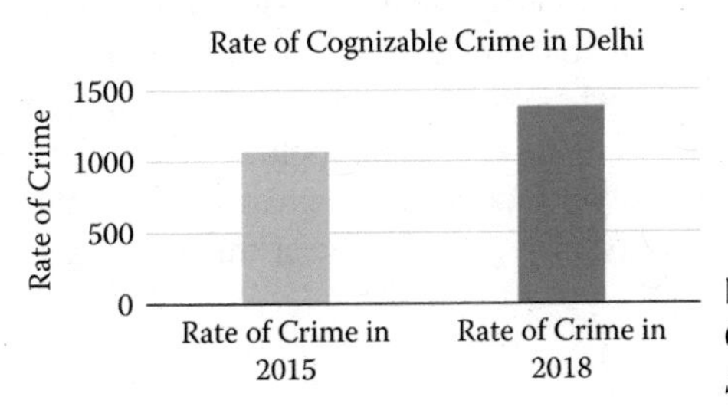

FIGURE 3.13 Graph Indicating Rate of Cognizable Crime in Delhi, India.
Source: NCRB (2015, 2018).

The incredible expansion of transport and communication facilities has rendered political boundaries notional and facilitates unfettered movement of criminals and criminal gangs across the region (Table 3.2). Some of the important factors impacting crime in Delhi are:

- The size and heterogeneous nature of its population.
- Wealthy income groups.
- Socioeconomic imbalances.
- Unplanned urbanization with a substantial population living in jhuggi-jhopri/kachchi colonies, etc., and the nagging lack of civic amenities therein.
- Proximity in location of colonies of the affluent and the under-privileged.
- A fast-paced life that breeds a general proclivity towards impatience, intolerance and high-handedness.
- Urban anonymity and slack family control.
- Easy accessibility/means of escape to criminal elements from across the borders.
- Extended hinterland in the NCR region.

Ahmad Uddin and Laxmi (2017) analyzed the crime data for 10 cities (highest crime rate) of India including all metro cities for the year 2015 to understand city crime trend towards various crimes types. By analyzing the crime data of 2015, the study

TABLE 3.2
Comparative Level of Crime in Million-Plus Cities

Attempt to Commit Murder Total: 4,697	Kidnapping and Abduction Total: 19,455	Robbery Total: 14,214
Delhi-674	Delhi-6,630	Delhi-6,766
Bengaluru-464	Mumbai-1,583	Mumbai-1,708
Mumbai-231	Patna-869	Pune-731

Source: Crime in India, NCRB (2015).

reveals that the crime density was in the range of 65.8 to 1,140, the lowest in Nagaland and highest in Delhi, which was found to be roughly 4.5 times than the national average. Delhi city reported highest in robbery and theft. After evaluation of crime percent change for the year 2014, it was found that 94.3% increase in crime in Delhi when compared with preceding year, whereas Mizoram got second position with value 25.2%. High incidents of crimes in mega cities/metropolitan cities may be due to various factors such as high density of population, greater information availability/flow, greater degree of anonymity in big cities, social milieu of urban slums, etc.

3.4.3.1 Crime Rate Reduction in the Future

In order to reduce the crime rate in the future, a comprehensive approach is required. It is essential to analyze the cause and effect of crime in the region. The future of the crime rate reduction could be dependent on identification of an ongoing crime and prevention in action. A fearful voice, a call for help, or a violent scene could be identified with the help of nano devices embedded with sirens. Around 70% of the crimes are committed by repeat offenders. If we are able to identify a previous offender and monitor closely, then the crime rate would drop significantly. It is not as simple as it sounds. Identifying previous antecedents is one of the toughest problems in India. There are information gaps everywhere, such as in the record keeping, digitization, access, accuracy, updation, and availability (FICCI, 2018).

3.4.3.2 Crimes Against Women

Time-series analysis is an important tool to determine presence of basic features of crime such as trends, seasonal behavior, etc. Time-series analysis correlates a series of observations collected at regular time intervals. A time series comprises four components (variations): trend variation, seasonal variation, cyclic changes, and irregular component. Dwivedi and Sandeep (2019) assessed the pattern of cognizable crimes committed against women in New Delhi in comparison to India as a whole in order to provide a comprehensive scenario using time-series analysis. A rising trend was noticed for crimes against women with the rate ranging from 204.6 (2016) to 308.8 (2019) per 1,00,000 women in New Delhi. The data showed that the population of women in New Delhi is 1.5% of that of India as a whole, while crime against women is slightly higher at 5.2%. In regard to the pattern of crime against women, it was noticed that the number of cases registered under outrage and insult to modesty was much higher in New Delhi (40.4%) in comparison to the country as a whole (27.8%). However, cruelty by husband and in-laws (members of husband's family) was less common in New Delhi (20.5%) in comparison to the whole country (34.6%). Similarly, kidnapping and abduction cases registered were also higher in New Delhi (25%) in comparison to the country (18.1%) (Table 3.3). There was a 40% increase in reported cases in New Delhi from 2009 to 2012 while the rise was 33% from 2013 to 2015. It was projected that if the current situation remains unchanged, there will be a 25% increase in crime against women.

The National Capital Region constitutes a mix of urban, slum, and urbanized rural population, with an 86% literacy rate along with one of the highest per capita income rates in India, but with a skewed sex ratio of 868 compared with 940 for the whole country in 2011. The cosmopolitan environment constitutes diverse economic, cultural, dietary, language, and religious practices. The porous boundaries of planned and unplanned urban development in the region is expanding and encroaching on neighboring states. Such increasing urbanization, material aspirations, and migration are leading to challenging pressures on space and basic utility services, increased pollution, and mushrooming of slum clusters. The proportion of cases registered as "cruelty by husband and relatives" was 20.5%, which appears to correlate with a slightly higher population-based prevalence of spousal violence (26.8%) experienced by women in New Delhi.

3.4.3.3 Measures to Control Crime in NCR

Measures such as sustained political commitment, increasing system accountability, social consciousness, digital awareness, removing online child pornography, restricting

TABLE 3.3
Crimes Against Women in Delhi, NCR (in %)

Crime	India	New Delhi
Outrage and insult to modesty	27.8	40.4
Kidnapping and abduction	18.1	25.0
Cruelty by husband and relatives	34.6	20.5
Rape and attempted rape	11.9	13.0
Dowry	5.4	0.3
Miscellaneous	2.2	0.8
Total	100	100

Source: Dwivedi & Sachdeva 2018.

migration, socioeconomic improvement, safe transport, gender sensitization training, counseling, surveillance, and increased crime control policing may lead to lower crime rates against women, elders, and children in society. In order to make use of the information technology, the government of India approved the design, development and implementation of a Government to Government (G2G) model called the Crime Criminal Information System (CCIS). Developed, under the aegis of the National Crime Record Bureau (NCRB), the Crime Criminal Information System (CCIS) is a G2G model designed to create computerized storage, analysis, and retrieval of crime criminal records (Gupta et al., 2006). Therefore, there is a need for integrated support systems (Chen et al., 2003; Amarnathan, 2003) as crime analysis tools based on current technologies to meet and fulfill the new emerging responsibilities and tasks of the police (Chaudhary, 2003).

Neighborhood Watch Scheme

In order to promote a long-lasting partnership with the community and to enhance safety and reduce the fear of crime, the mechanism of Neighborhood Watch Scheme is one of the successful initiatives by the Delhi police in the city.

Eyes and Ears Scheme

In the "Eyes and Ears" scheme, various sections of the public, such as *rehriwalas, chowkidars, patriwalas,* security guards, parking attendants, three wheeler/taxi drivers, bus drivers/conductors, porters, shopkeepers, property agents, second-hand car dealers, landlords, members of RWA/MWA, cyber café owners, PCO owners, guest house owners, and other alert citizens are involved in providing information regarding suspicious activities of individuals and crimes.

Concluding Remarks

Due to size, location, demography, and resources, India has emerged as an emerging powerful country not only at a regional level but also playing an important role in the world's economy and political arena. Inspite of all these achievements, the country is facing various kind of problems within as well as across the borders. Insurgency, cross-border terrorism, border disputes with neighboring countries, infiltration, and refugee movement are some of the important issues a hunting the country for years. It has become now intertwined with organized crime, human trafficking, and corruption. In ancient India, women were revered and rendered a significant contribution towards family and society but as of today, the position of women in Indian society is saddening. Ever-increasing numbers of youth involved in heinous crimes are making the country bleed internally, which may have serious consequences for the country and society as a whole.

REFERENCES

Agarwal, D. (2018) Juvenile delinquency in India- Latest trends and entailing amendments in Juvenile Justice Act. *People* 3(3).

Ahmad, F., Md Meraj Uddin, and Laxmi, G. (2017) Role of Geospatial Technology in Crime Mapping: A Perspective View of India. World Scientific News. http://www.worldscientificnews.com/wp-content/uploads/2017/08/WSN-882-2017-211-226.pdf.

Amarnathan, L. C. (2003) Technological advancement: Implications for crime, The Indian Police Journal, April-June. In Gupta et al., (2006) *Crime Data Mining for Indian Police Information System*. Research Gate. https://www.researchgate.net/publication/255512466.

Burt, C. (1925) *The Young Delinquent*. University of London Press. In Venkatachalam, C. and Aravindan, S. (2014) Juvenile delinquency as a result of broken homes. *J. Inter. Acad. Res. Multidisc.*, 2(8). http://www.jiarm.com/SEP2014/paper17462.pdf.

Chen, H., Zeng, D., Atabakhsh, H., Wyzga, W., and Schroeder, J. (2003) COPLINK: Managing Law Enforcement Data and Knowledge. Communications of the ACM 46 (1), 28–34. In Gupta et al. (2006) *Crime Data Mining for Indian Police Information System*. Research Gate.

Chaudhary, J. N. (2003) Police in United States: Contemporary Issues. The Indian Police Journal, July–September, 173–179. In Gupta et al. (2006) *Crime Data Mining for Indian Police Information System*. Research Gate.

Choudhary, S. (2017) Juvenile delinquency: Elementary concepts, causes and prevention, *Inter. J. Hum. Soc. Sci. Res.*, 3(5), 52–59. file:///C:/Users/dell%20vostro/Downloads/3-5-26-493%20(2).pdf.

Cities Alliance. (2015) Research Paper: Managing Peri-Urban Expansion. Cities Alliance Project Output, Knowledge Support for PEARL Programme under JNNURM. ocuments.worldbank.org/curated/en/755241467991968684/pdf/99277-WP-P121456-Box393195B-PUBLIC-research-paper.

Dreze, J. and Reetika, K. (2000) Crime, gender, and society in India: Insights from homicide data. *Pop. Dev. Rev.*, 26(2) pp. 335–352. http://www.jstor.org/stable/172520.

Ease of Living Index. (2018) SMARTNET, Ministry of Housing and Urban Affairs, Govt. of India. https://www.ipsos.com/sites/default/files/ct/publication/documents/2018-08/ease-of-living-national-report.pdf.

Ford, J. A. (2005) Substance use, the social bond, and delinquency. *Sociol Inq.*, 7, 109–128. In Mhavan, N. (2017) Juvenile delinquency: A social perspective. *Schol. Res. J. Hum. Sci. Eng. Lang.*, 6/28. http://oaji.net/articles/2017/1201-1537182511.pdf.

Farhadian, A. (2016) Study of factors affecting delinquency of children. International Journal of Humanities and Cultural Studies, 2(4), p. 1907. https://pdfs.semanticscholar.org/c198/a786ffb8277f10d6e19b8ca91930e72f4bdb.pdf.

FICCI. (2018) India Risk Survey (IRS) 2018. Pinkerton Consulting & Investigations, Inc and Federation of Indian Chambers of Commerce and Industry (FICCI). http://ficci.in/Sedocument/20450/India%20Risk%20Survey%20-%202018.pdf.

Ghosh, D. (2013) "100 Juveniles Booked for Murder, 63 Held for Rape," The Times of India. June 14, p 4: In Agarwal, D. (2018). Juvenile Delinquency In India- Latest Trends And Entailing Amendments In Juvenile Justice Act. People: International Journal Of Social Sciences, 3(3), 1365–1383. https://www.researchgate.net/publication/322918203.

Global Extremism Monitor. 2017. https://institute.global/sites/default/files/inline-files/Global%20Extremism%20Monitor%202017.pdf.

Gottfredson, M. R. and Hirschi, T. (1990) *A General Theory of Crime*. Stanford: Stanford University Press. In Mhavan, N. (2017) Juvenile delinquency: A social perspective, *Schol. Res. J. Hum. Sci. Eng. Lang.*, 6/28. http://oaji.net/articles/2017/1201-1537182511.pdf.

Gupta, M., Chandra, B., and Gupta, M. P. (2006) Crime Data Mining for Indian Police Information System. Research Gate. https://www.researchgate.net/publication/255512466.

Hao, M., D., Jiang, Ding, F., Jingying, Fu., and Shuai, C. (2019) Simulating Spatio-Temporal Patterns of Terrorism Incidents on the Indochina Peninsula with GIS and the Random Forest Method. https://doi.org/10.3390/ijgi8030133.

Healy, W. and Healy, M. T. (1917) Pathological lying, accusation and swindling; A study in forensic psychology. Little, Brown & Co. In Venkatachalam, C. and Aravindan, S. (2014) Juvenile delinquency as a result of broken homes. *J. Inter. Acad. Res. Multidiscip.*, 2(8). http://www.jiarm.com/SEP2014/paper17462.pdf.

IEP. (2019) Global Terrorism Index 2019 Measuring the Impact of Terrorism. The Institute for Economics & Peace (IEP). http://visionofhumanity.org/app/uploads/2019/11/GTI-2019web.pdf.

James, L. C. (1982) Significant Refugee Crises Since World War II and the Response of the International Community. *Michigan J. Inter. Law*, 3, 1. https://repository.law.umich.edu/cgi/viewcontent.cgi?article=1859&context=mjil.

Knoester, C. and Haynie D. L. (2005) Community context, social integration into family, and youth violence. *J. Marr. Fami.*, 67(3), 767–780. https://doi.org/10.1111/j.1741-3737.2005.00168.x.

Kumar, B. B. (2006) Introduction. In Kumar B. B. and Astha B. (Eds.) *Illegal Migration from Bangladesh* (pp. 1–8). New Delhi: Concept Publishing Company.

Leeper, R. D. (1925) A study of juvenile delinquency in thirty counties in Idaho. *J. Crim. Law Criminol.*, 16, 388–436. In Venkatachalam, C. and Aravindan, S. (2014) Juvenile delinquency as a result of broken homes. Journal of International Academic Research for Multidisciplinary, 2(8). http://www.jiarm.com/SEP2014/paper17462.pdf.

Marte, R. M. (2008) Adolescent Problem Behaviour: Delinquency, Aggression, and Drug Use. New York: LFB Scholarly Publishing, LLC. In Ntshangase, M. P. (2015) A Study of Juvenile Delinquency Amongst Adolescents in Secondary Schools in Gauteng, Submitted in fulfilment of the requirements for the degree of Master of Education, University of South Africa. https://pdfs.semanticscholar.org/6568/b589d4dcf8f3ec71e85b0b3507dd251b26c4.pdf.

Matza, D. (1964) *Delinquency and drift. Social Forces.* Vol. 43(4) John Wiley and Sons. https://doi.org/10.2307/2574506.

MHA. (2018) Terrorist Infiltration Into J-K In 2018 Highest in Five Years: MHA Report. The Economic Times (October, 26, 2019). E-Paper. https://economictimes.indiatimes.com/news/defence/terrorist-infiltration-into-j-k-in-2018-highest-in-five-years-mha-report/articleshow/71774332.cms.

Mhavan, N. (2017) Juvenile delinquency: A social perspective. *Schol. Res. J. Human. Sci. English Lang.*, 6/28. http://oaji.net/articles/2017/1201-1537182511.pdf.

Mishra, A. K. N. (2013) Juvenile Delinquency on the Rise. Times of India. July 13. In Agarwal, D. (2018) Juvenile delinquency in India—Latest trends and entailing amendments in Juvenile Justice Act. *People* 3(3), 1365–1383. https://www.researchgate.net/publication/322918203.

Miller, W. B. (1958) Lower class sub-culture as a generating mileu of gang delinquency. *J. Soc.*, 14, 5–19.

NCRB. (2015) Crime in India 2015, Compendium. National Crime Record Bureau. http://ncrb.gov.in/StatPublications/CII/CII2015/FILES/Compendium-15.11.16.pdf.

NCRPB. (1999) National Capital Region Growth and Development, National Capital Region Planning Board, Har-Anand Publications Pvt. Ltd. http://ncrpb.nic.in/pdf_files/Growth&Development.PDF.

NCRPB. (2018) Annual Report, 2017–18. National Capital Region Planning Board Ministry of Housing and Urban Affairs, Government of India. http://ncrpb.nic.in/pdf_files/annualreport2017-18.pdf.

Putwain, D. and Sammons, A. (2002) Psychology and Crime. Psychology Press. In Farhadian, A. (2016) Study of factors affecting delinquency of children. *Inter. J. Human. Cult. Stud.*, 2(4). https://pdfs.semanticscholar.org/c198/a786ffb8277f10d6e19b8ca91930e72f4bdb.pdf.

Saji, C. *Terrorism and Legal Policy in India – An Overview.* https://satp.org/satportp Publication.

Saikia, A. (2005) Refugees, Illegal Migrants and Local Perceptions in India's Far East. Saikia N., Saha A., Bora J. and Joe W. (2016) Trends in immigration from Bangladesh to Assam, 1951–2001: Evidence from Direct and Indirect Demographic Estimation IMDS Working paper Series. https://www.researchgate.net/publication/322222991.

Sanajai, S. (2005) Strategic Perspectives: Border Management of Western Borders of India. http://www.usiofindia.org/pers/.htm.

Shaw, C. R. and McKay, H. D. (1931) Social Factors in Juvenile Delinquency. Report on the Causes of Crime. Vol. 2. National Commission on Law Observance and Enforcement. Washington, DC. https://www.ncjrs.gov/pdffiles1/Digitization/44551NCJRS.pdf.

Shoemaker, D. J. (2009) *Juvenile Delinquency*. Lanham, MD: Rawson & Littlefield Publishers. https://epdf.pub/juvenile-delinquency.html.

Tata Trusts. (2019) India Justice Report, Ranking States on Police, Judiciary, Prisons & Legal Aid. https://www.tatatrusts.org/upload/pdf/overall-report-single.pdf.

UNHCR. (2019) Worldwide Displacement Tops 70 Million, UN Refugee Chief Urges Greater Solidarity in Response. The UN Refugee Agency, UNHCR. https://www.unhcr.org/news/press/2019/6/5d03b22b4/worldwide-displacement-tops-70-million-un-refugee-chief-urges-greater-solidarity.html.

USDSP. (2019) Country Reports on Terrorism, 2018. United States Department of State Publication (USDSP). https://www.state.gov/country-reports-on-terrorism-2/.

Upadhyaya, R. P. (2017) Crime in Delhi. http://www.delhipolice.nic.in/ANNUAL%20REVIEW_2017.pdf. https://www.indiatoday.in/diu/story/india-saw-almost-1-500-acid-attacks-in-five-years-1636109-2020-01-12.

Venkatachalam, C. and Aravindan, S. (2014) Juvenile delinquency as a result of broken homes. *J. Inter. Acad. Res. Multidiscipli.*, 2(8), p. 473. http://www.jiarm.com/SEP2014/paper17462.pdf.

4 Identifying Crime Hot Spots

The growth of intelligence-led policing has placed a greater emphasis on the effective identification of crime hot spots as well as the choice of the crime reduction or detection strategy identified to combat a problem. Crime hot spots have become strategically important locations for policing surveillance and functioning in many locations, as they enable an operational commander to focus resources into the areas of highest need. The chapter outlines the techniques used to identify the spatial and temporal components of crime hot spots, and utilizes these methods to identify three broad categories of temporal hot spots and three broad categories of spatial hot spots. Real examples show how through spatial interpolation, inverse distance, weight, age, and kriging, combined within the regression analysis, yield operational results for appropriate crime prevention or detection strategy.

4.1 SPATIAL INTERPOLATION AND GIS

Spatial analysis is the process of using spatial information to extract new information and meaning from the original data. The interpolation process uses points with known values to estimate values at other unknown points. GIS uses spatial analysis tools for calculating feature statistics and carrying out geoprocessing activities as data interpolation. This tool is immensely applicable in research for terrain analysis, slope profiling, and hydrological modeling (modeling the movement of water over and in the earth). In wildlife management, it can be used for analytical functions dealing with wildlife point locations and their relationship to the environment. In the field of climatology, spatial analysis is used for making precipitation (rainfall) maps for a country; to estimate the temperature variations at different locations without recorded data by using the known temperature readings from nearby weather stations. Such interpolated surfaces are often known as statistical surfaces. This tool can be used to predict unknown values for any geographic point data, such as chemical concentrations, noise levels, and snow accumulation; water table and population density are other similar types. Large data collection sometimes seem to be a herculean task that is cost intensive and cumbersome as well. In such cases, data collection is usually conducted only in a limited number of selected point locations. Later, with the help of software, spatial interpolation of these points is used to create a raster surface with estimates made for all raster cells (Fig. 4.1).

In order to create a continuous map for digital elevation from elevation points measured with a GPS device, a suitable interpolation method has to be used to

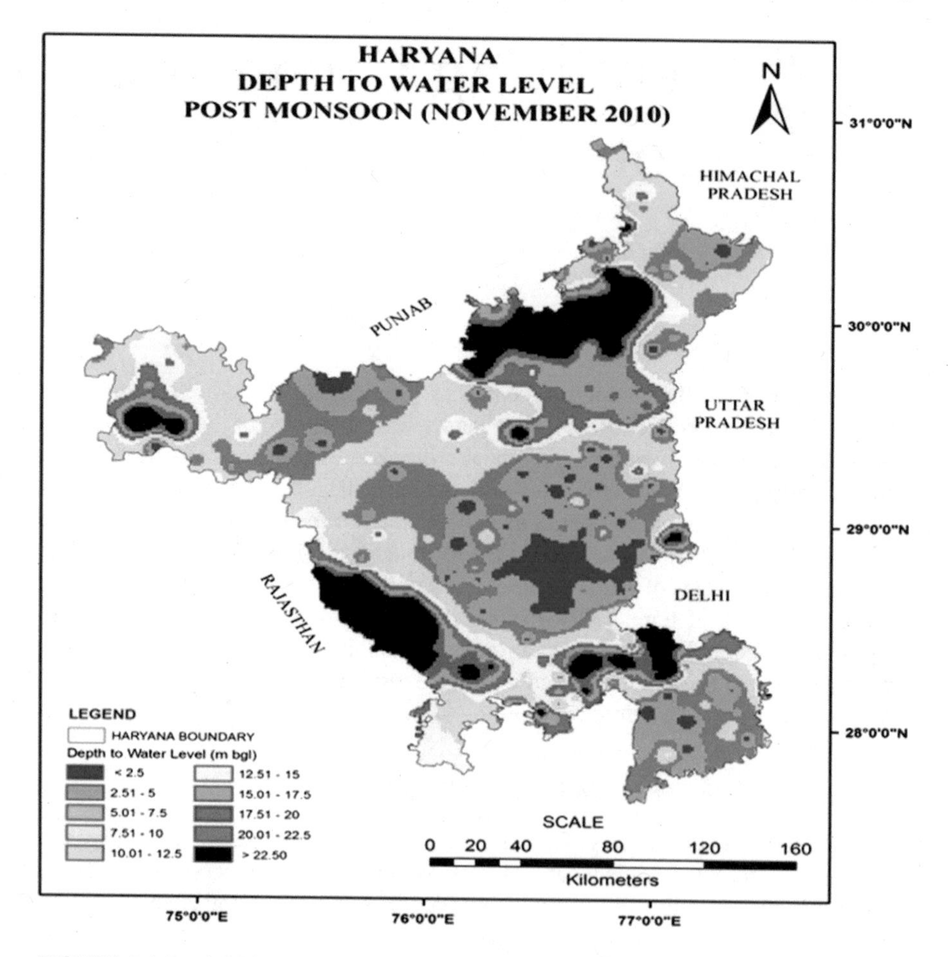

FIGURE 4.1 Spatial Distribution Map of Haryana (India) Showing Depth to Water Level.

optimally estimate the values at those locations where no samples or measurements were taken. The results of such interpolation analysis can then be used for analyses that cover the whole area for modeling.

The use of spatial interpolation in crime studies is very impactful. The crime locations, their frequency of occurrences, and geographical understanding of a region, together lay emphasis on its analysis. Spatial distribution of incidents include many analyses such as the center of minimum distance, standard deviational ellipse, and the convex hull and directional mean.

Standard deviational ellipse involves a common way of measuring the trend for a set of points or areas to calculate the standard distance separately in the x and y directions. These two measures define the axes of an ellipse encompassing the distribution of features in a region. The ellipse is referred to as the standard deviational ellipse, since it calculates the standard deviation of the x coordinates and y coordinates from the mean center to define the axes of the ellipse. The ellipse allows you to observe if the distribution of features is elongated and hence has a particular orientation (Fig. 4.2).

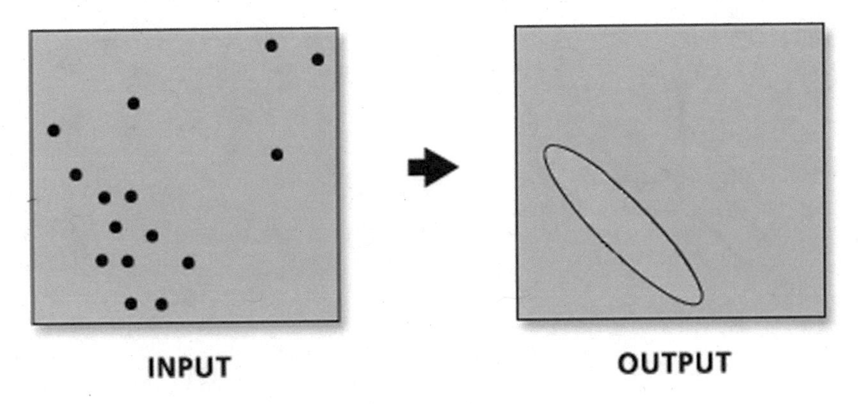

FIGURE 4.2 Distribution Trend with Standard Deviational Ellipse.

The investigator can interpret a sense of orientation by drawing the features on a map, calculating the standard deviational ellipse to make the trend clear. One can also calculate the standard deviational ellipse using either the locations of the features or using the locations influenced by an attribute value associated with the features. The latter is termed a "weighted standard deviational ellipse." The potential application of this tool is for mapping the distributional trend for a set of crimes and help identifying its relationship to a particular physical feature (a string of bars or restaurants, a particular boulevard, and so on).

It can also be used for mapping groundwater well samples for a particular contaminant and might indicate how the toxin is spreading and, consequently, may be useful in deploying mitigation strategies. It helps compare the size, shape, and overlap of ellipses for various racial or ethnic groups and may provide insights regarding racial or ethnic segregation. Plotting ellipses for a disease outbreak over time may be used to model its spread effectively, using this tool.

Convex hull in geometry represents a closure of a shape that contains the convex. It can be better understood as encircling the crime incidence in a closet and analyzing it (Fig. 4.3).

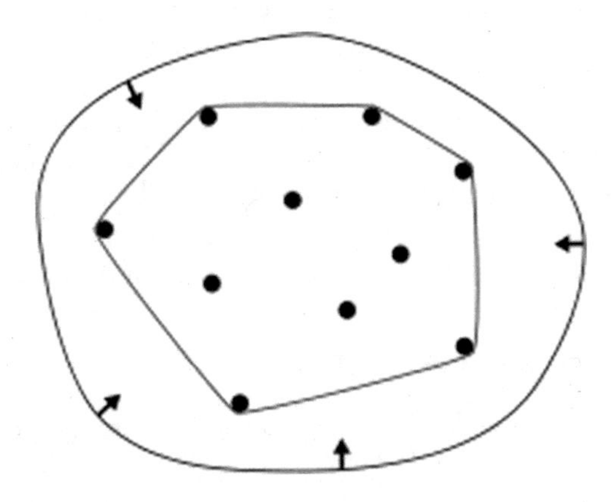

FIGURE 4.3 Convex Hull.

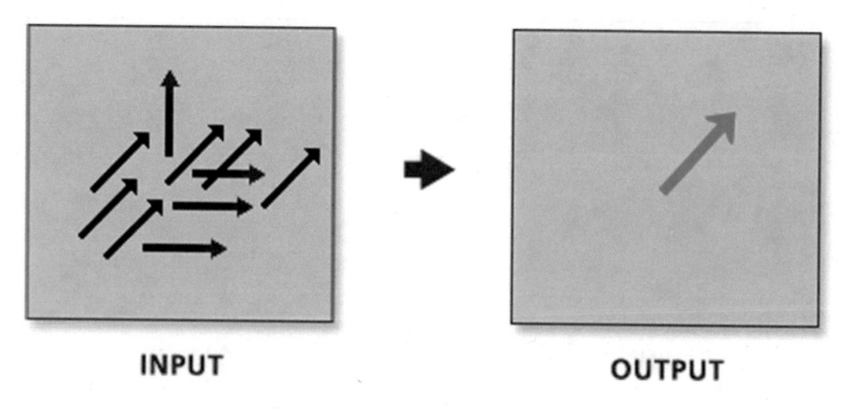

FIGURE 4.4 Directional Mean.

Directional mean is something which is related to or indicates the direction in which someone is moving. In particular it indicates the progression direction, which is very helpful for police patrolling and tracking culprits (Fig. 4.4).

The linear directional mean tool is used to calculate the trend of either the direction or orientation of line features by calculating the average angle of the lines. This statistic is used to evaluate auto theft data that contains information on the location of occurrence from where each vehicle was taken and where it was eventually recovered. The linear directional mean tool facilitates highlighting the recurring patterns that can suggest an underlying infrastructure supporting car thefts in the region. Similar analysis has been used to study data on missing or abducted children in a region.

When crime distributions are compared to other features in the landscape, similarities or relationships often become evident. The most common way for measuring the trend for points or areas is to calculate the standard distances separately in x and y directions. These two measures define the axis of an ellipse encompassing the distribution of features. The ellipse is referred to as the standard deviational ellipse since the method calculates the standard deviation of the x-coordinates and y-coordinates from the mean center to define the axis of the ellipse. This ellipse shows if the distribution of features is elongated, it has a particular orientation.

In some cases, crime events grouped by police beat and evaluated using the standard deviational elipse tool may indicate that, in some police beats, crime activities were evenly distributed throughout the beat, thus the ellipse resembles a circle. Whereas, in other cases, crime activities tend to follow some road networks and crime incidents in these police beats highlight that orientation (Fig. 4.5).

4.1.1 Inverse Distance Weight (IDW)

The Inverse Distance Weight (IDW) is a popular interpolation technique which considers that each input point has a local influence that diminishes with increasing distance. It weighs the points closer to the processing cell greater than those comparatively farther away (Fig. 4.6). For an analysis with IDW

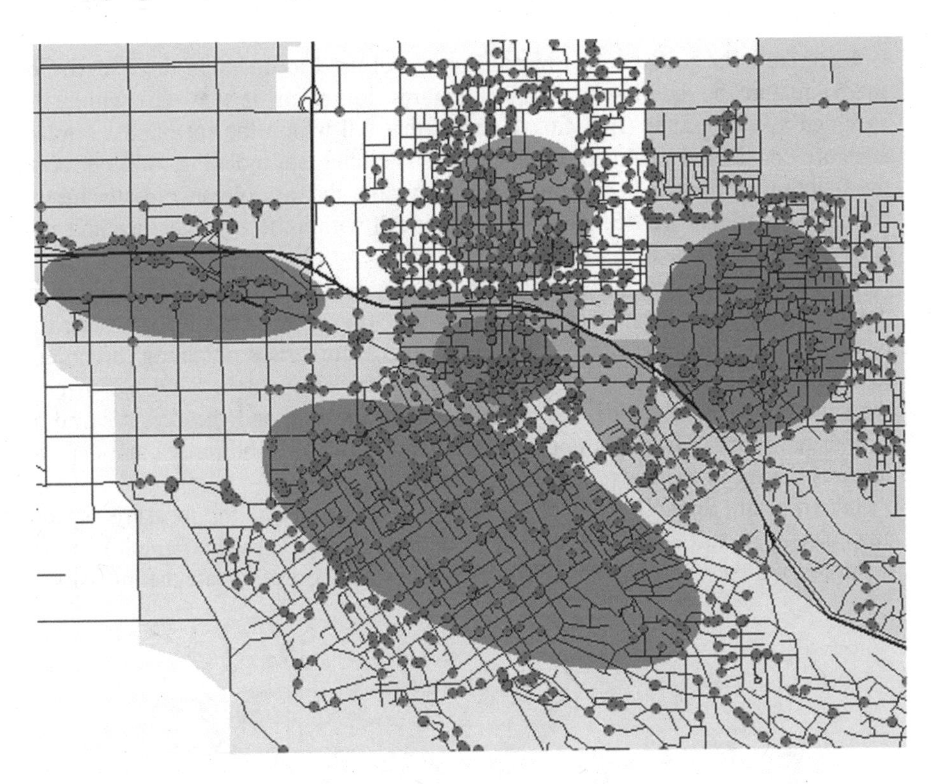

FIGURE 4.5 Analysis of the Spatial Dispersion of Crimes by Police Beat in Redlands, California. This Shows That Crimes in the Western Portion of the City Follows Major Transportation Network Routes in the Area.

either a specified number of points or all points within a given radius can be used to determine the output value of each location. Use of this method assumes the variable being mapped decreases in influence with distance from its sampled location.

The IDW algorithm effectively is a moving average interpolator that is generally applied to highly variable data. For certain data types, it is possible to return to the collection site and record a new value that is statistically different from the original reading but within the general trend for the area. The interpolated surface, estimated using a moving average technique, is less than the local maximum value and greater than the local minimum value.

The Fig. 4.10 discusses the use of IDW in detail, highlighting the crime-prone zones of Ajmer City, Rajasthan (India). A study under a major project on impact of geographical space and urban transformation on women in society: a study of Ajmer City, Rajasthan (India), sponsored by ICSSR (IMPRESS Scheme), Government of India. To initiate the process, the point file of all registered crimes of Ajmer City for a certain year was created and the integrate tool was applied with a tolerance level of 30 meters.

Integrate is basically a data management tool used to maintain the integrity of shared feature boundaries by making features coincident if they fall within the specified x, y tolerance (Fig. 4.6). Features that fall within the specified x,y tolerance are considered identical or coincident. The integrate tool (Fig. 4.7) performs the function of finding the features that are within the x,y tolerance. Later inserts common coordinate vertices for features that fall within the x,y tolerance and will add vertices where features intersect.

Later, the event data (crime locations in this case) is converted into weighted point data by applying the collect events tool (Fig. 4.8). This feature basically combines the coincident points, creating a new feature class obtaining the unique locations found in the Input Feature Class. It then adds a field ICOUNT to hold the sum of all crime incidences at each unique location and the file is saved as aggregated points. Here it is important to mention that this tool only combines the features that have the exact same coordinates.

Incremental Spatial Autocorrelation is further used to measure spatial autocorrelation for a series of distances, optionally creating a line graph of those distances and their corresponding Z-scores. These Z-scores indicate the intensity of

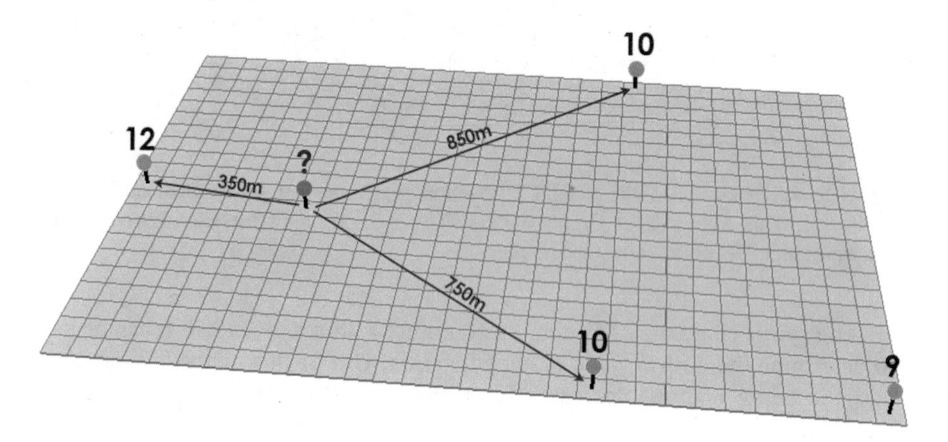

FIGURE 4.6 Decrease in Value with Increasing Distance from Specified Location.

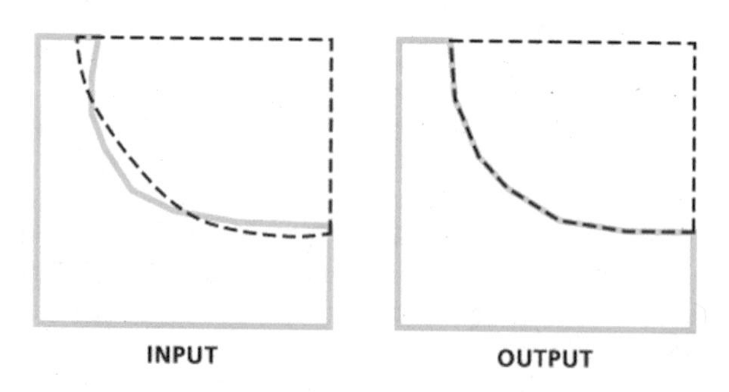

FIGURE 4.7 Results Observed after Applying Integrate Tool.

spatial clustering of crimes and statistically significant peak. Z-scores reflect the distances where spatial processes promoting clustering are most pronounced. Taking the collect event file as an input feature and input field as ICOUNT, the Z-score value hot spot report is obtained.

The hot spot analysis tool calculates the Getis-Ord Gi* statistic (pronounced G-i-star) for each feature in a data set now before finally applying the IDW. The resultant Z-scores and P-values helps us understand where features with either high or low values cluster spatially exist. The aggregated crime point file is taken as input feature class, with ICOUNT as the input field with the distance band or threshold as the highest Z-score value as observed in the report generated during the Incremental Spatial Autocorrelation process (Encyclopedia of GIS, 2008). Thus an aggregated hot spot file is created with GiZscore. The Gi* statistic returned for each feature in the dataset is a Z-score (Fig. 4.9). For statistically significant positive Z-scores, the

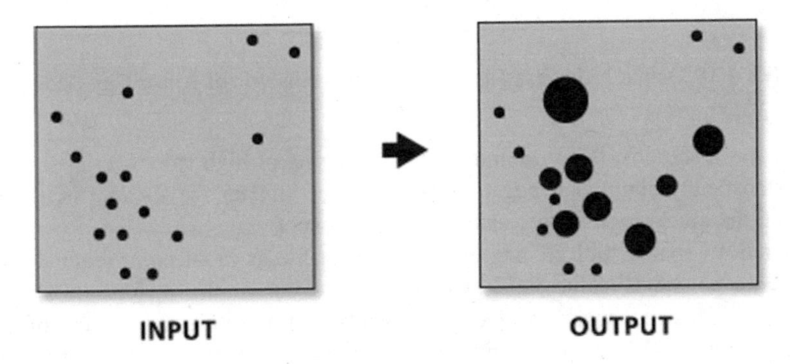

FIGURE 4.8 Results Observed after Applying Collect Events Tool.

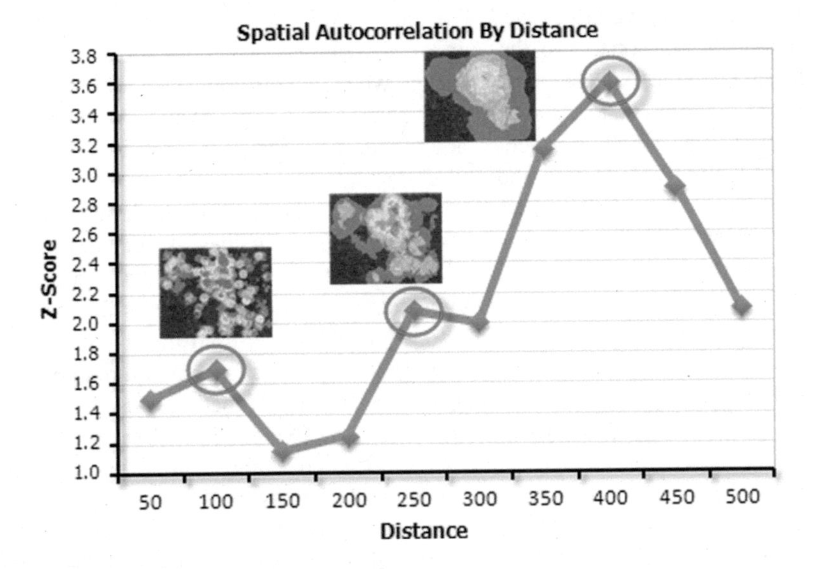

FIGURE 4.9 Example of Line Graph of Z-Scores.

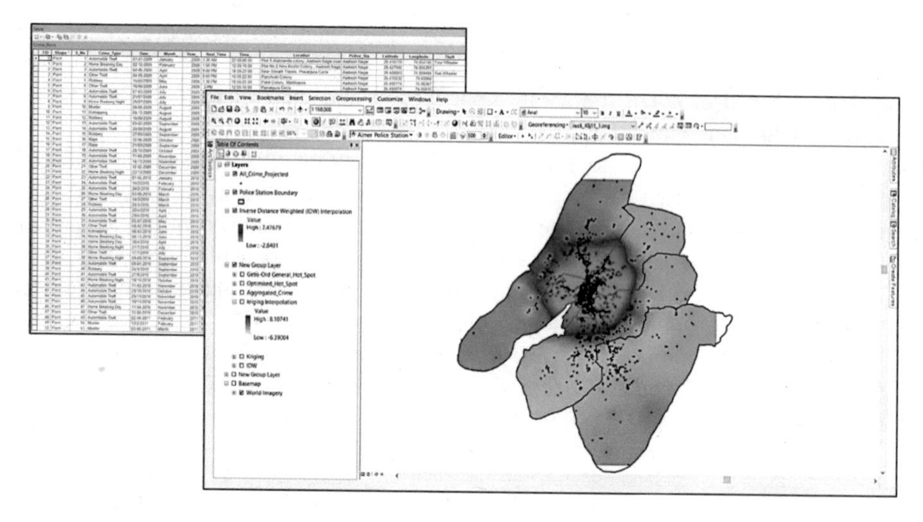

FIGURE 4.10 Output of Inverse Distance Weighted Analysis of Ajmer City (India).

larger the Z-score is, the more intense the clustering of high values (crime hot spot). For statistically significant negative Z-scores, the smaller the Z-score is, the more intense the clustering of low values (crime cold spot).

For IDW spatial tool the aggregated hot spot file as input point feature, use the GiZscore as the Z value field and apply it to obtain the results, as shown in Fig. 4.10. This tool works by looking at each feature within the context of neighboring features. We need to understand here that a feature with a high value is interesting but may not be a statistically significant hot spot. To be a statistically significant hot spot, a feature should have a high value and be surrounded by other features with high values as well. The local sum for a feature and its neighbors are compared proportionally to the sum of all features; when the local sum is very different from the expected local sum, and when that difference is too large to be the result of random chance, a statistically significant Z-score results. When the False Discovery Rate (FDR) correction is applied, statistical significance is adjusted to account for multiple testing and spatial dependency.

The shade around the core crime region is indicating the highest values with maximum weights, whereas as we move away, the values are decreasing. IDW interpolation explicitly implements the assumption that things that are comparatively closer to one another are more alike than those that are further apart. To predict a value for any unmeasured location, IDW has used the measured values around the prediction location. Thus, IDW presumes that each measured point has a local influence that diminishes with distance. The IDW function is used here for the set of points and is dense enough to capture the extent of local surface variation which is needed for analysis. IDW determines cell values using a linear-weighted combination set of sample points. It weighs the points closer to the prediction location greater than those farther away, hence the name inverse distance weight (Rocha et al., 2018). An IDW technique value for each grid node has been calculated by examining surrounding data points that lie within a user-defined search radius. Some or all of the data points are used in the interpolation process for analysis. The node value is calculated by averaging the weighted sum of all the points. Data points that lie progressively farther from the node influence the computed value far less than those lying closer to the node. A radius is generated around each grid node from which data points are selected to be used in the calculation. Options to control the use of IDW include power, search radius, fixed search radius, variable search radius, and barrier.

4.1.2 Kriging

Kriging is a geostatistical interpolation technique that considers both the distance and the degree of variation between known data points while estimating values in unknown areas (Rocha et al., 2018). It is a multistep process; it includes exploratory statistical analysis of the data, variogram modeling, creating the surface, and (optionally) exploring a variance surface. Kriging is basically a statistical method to calculate weights based on neighboring values. It is also known as BLUE (Best Linear Unbiased Estimator).

For ordinary kriging (with local mean), all crime locations within the police boundary were taken as a source data set using a fish net. The basic idea using kriging is to predict the value of a function at a given point. Ordinary kriging helps generate probability maps for crime incidences based on incident records showing that regions in the northwest and central city are more susceptible to crime incidences (Fig. 4.11). A gradual decrease of probability of occurrence in the neighboring wards was observed with low values in the periphery region.

Co-kriging uses crime incidences as the primary variable, along with the socio-economic indicators with strong co-relations with the primary variable for future predictions. As co-kriging requires much more estimation, including estimating the correlation for each dependent variable with the independent variable, it uses information on several variable types. Here illiterate population, non-working population, and sex ratio were taken as the dependent variables and overall crime incidences for the year 2011 were taken as the independent variable. Results indicated a very high positive correlation was observed between crime and the illiterate and non-working population. The co-kriging technique results in a formation of three "high-crime susceptible regions" in the city area within the police boundary (Fig. 4.12).

A kriged estimate is a weighted linear combination of the known sample values around the point to be estimated (Bhattacharjee, 2019). The kriging process generates an estimated surface from a scattered set of points with Z-values. Kriging assumes that the distance or direction between sample points reflects a spatial correlation that can be used to explain variation in any surface. The kriging tool fits a mathematical function to a specified number of points, or to all the points within a specified radius, to determine the output value for each location. This technique is most appropriate when there are spatially correlated distances or directional bias in the data. It is often used in soil science and geology (Bhattacharjee, 2019).

The above technique is applied on several points with the same x, y coordinates. If the values of the points at the common location are the same, they are considered

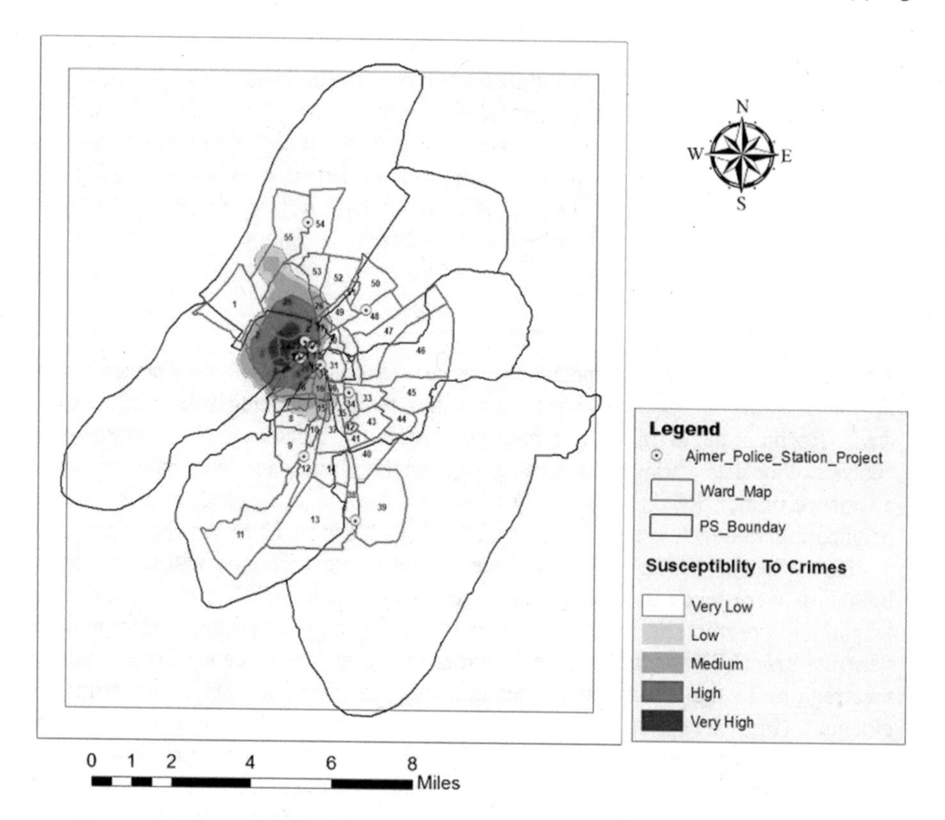

FIGURE 4.11 Crime Susceptibility of Ajmer City (India) Using Ordinary Kriging.

duplicates and have no effect on the output. If the values are different, they are considered coincident points.

The various interpolation tools may handle this data condition differently. For example, in some cases, the first coincident point encountered is used for the calculation; in other cases, the last point encountered may be used. This may cause some locations in the output raster to have different values than what you might expect. The solution is to prepare your data by removing these coincident points. The Collect Events tool in the Spatial Statistics toolbox is useful for identifying any coincident points in the data before applying kriging. Fig. 4.13 displays another example of the kriging tool application.

4.1.3 Density

The Density tools produce a surface that represents how much or how many of some things are there in per unit area. The Density tool is useful to create density surfaces to represent the distribution of crime incidents from a set of observations, or the degree of urbanization of an area based on the density of roads. In the density maps, a circular search area is applied that determines the distance to search for

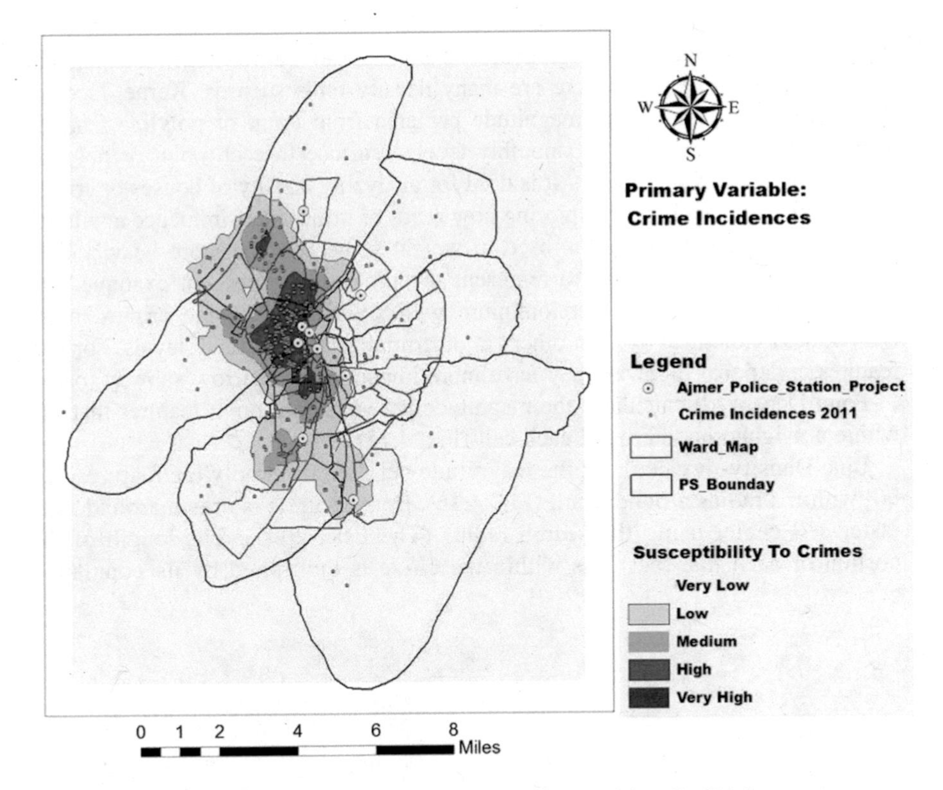

FIGURE 4.12 Crime Susceptibility of Ajmer City (India) Using Co-Kriging.

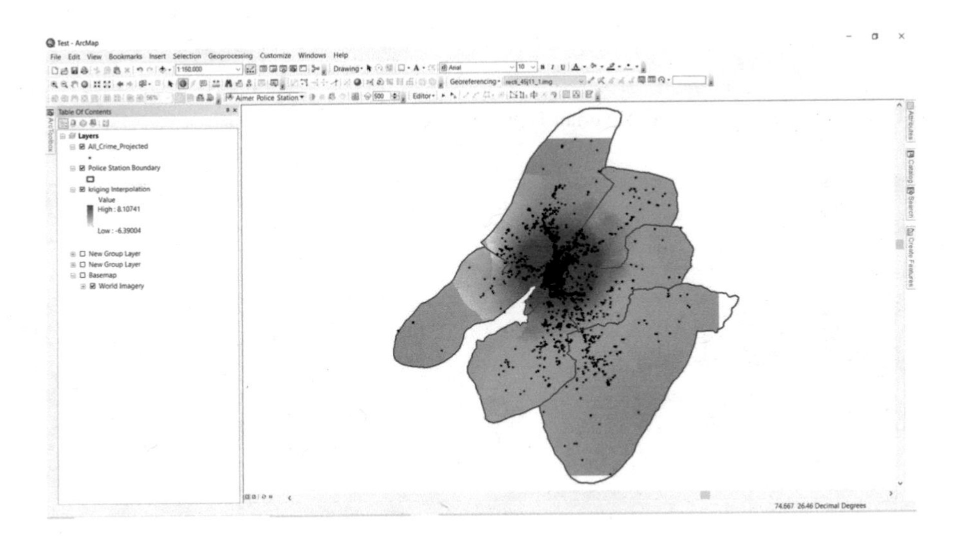

FIGURE 4.13 Example of Kriging Tool Application in Ajmer City (India).

sample locations (line or point) or to spread the values out around each location and calculate a density value. There are many density tools such as, Kernel Density (Fig. 4.14); it calculates the magnitude per area from point or polyline features using a kernel function to fit smoothly tapered surface to each point or polyline (Geospatial Intelligence, 2019). It is used for analyzing density of houses or crimes for community planning, or exploring how roads or utility lines influence a wildlife habitat. The population field is used to weight some features more heavily than others, or to allow one point to represent several observations. For example, one address might represent a condominium with six units, or some crimes might be weighted more heavily than others in determining overall crime levels. For line features, a national highway may have more impact than a narrow subway road.

Point Density- It calculates the magnitude per area from point features that fall within a neighborhood around each cell (Fig. 4.15).

Line Density- It calculates the magnitude per area from polyline features that fall within a radius around a cell (Fig. 4.16). Here, a circle is drawn around each raster cell center using the search radius (The Esri, 2005). The length of the portion of each line that falls within the circle is multiplied by its population

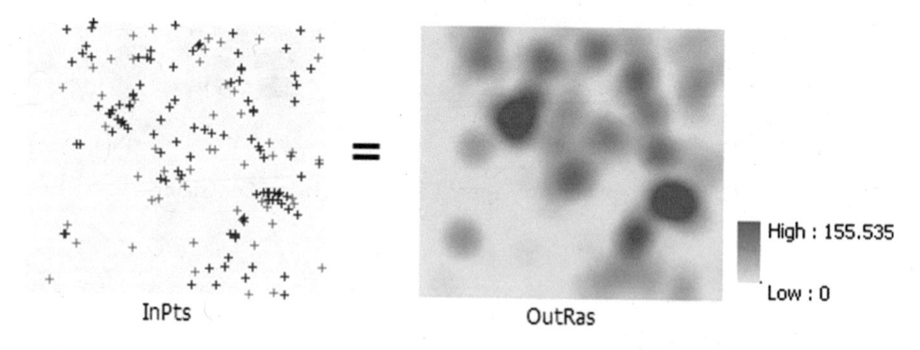

FIGURE 4.14 Application of Kernel Density Tool.

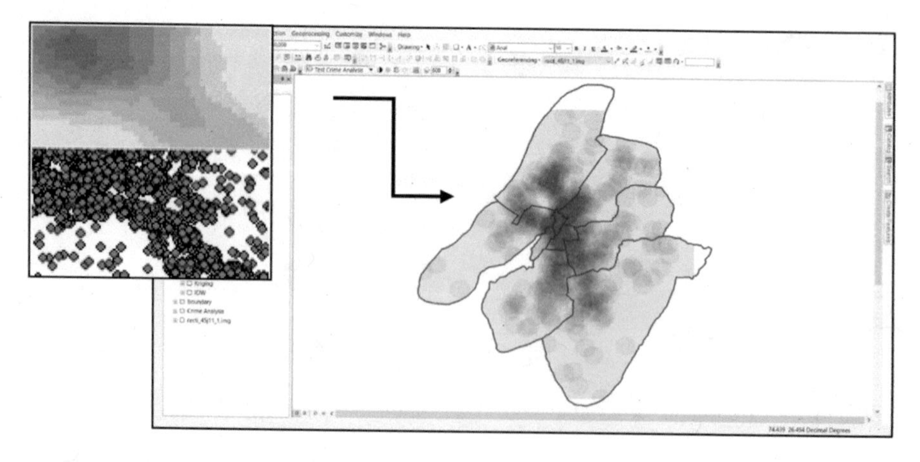

FIGURE 4.15 Application of Density Tool (Point Density) in Ajmer City (India).

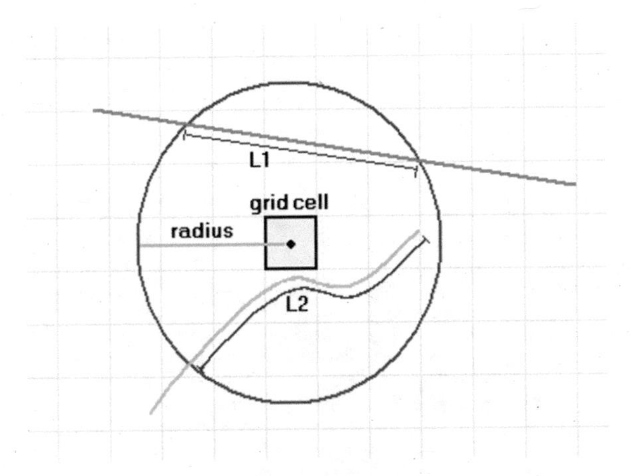

FIGURE 4.16 Line Density Tool.

field value. Then figures are summed, and the total is divided by the circle's area.

A raster cell and the circular neighborhood is used to determine the length for the line density. In Fig. 4.16, a raster cell is shown with its circular neighborhood. Lines L1 and L2 represent the length of the portion of each line that falls within the circle. The corresponding population field values are V1 and V2. Thus:

$$\text{Density} = ((\text{L1} \times \text{V1}) + (\text{L2} \times \text{V2}))/(\text{Area of Circle})$$

If a population field other than NONE is used, the length of the line is considered to be its actual length times the value of the population field for that line. Fig. 4.17 projects the use of the line density tool for minor and major road transport network in Ajmer City, Rajsathan, India.

4.2 HOT SPOT ANALYSIS

Hot spot analysis has been already discussed in brief with the IDW tool earlier. However, this section will help better understand the process and applications of hot spot analysis in details. This tool uses vectors to identify locations of statistically significant hot spots and cold spots in the data by aggregating points of occurrence into polygons or converging points that are in proximity to one another based on a calculated distance. The analysis groups feature when similar high (hot) or low (cold) values are found in a cluster. The polygons usually represent administration boundaries or a custom grid structure.

Prior to applying hot spot analysis, there is a need to test for the presence of clustering in the data with some prior analysis technique involving spatial autocorrelation to help identify if any clustering occurs within the entire dataset. Two available methods are Moran's *I* (Global) and Getis-Ord General *G* (Global)

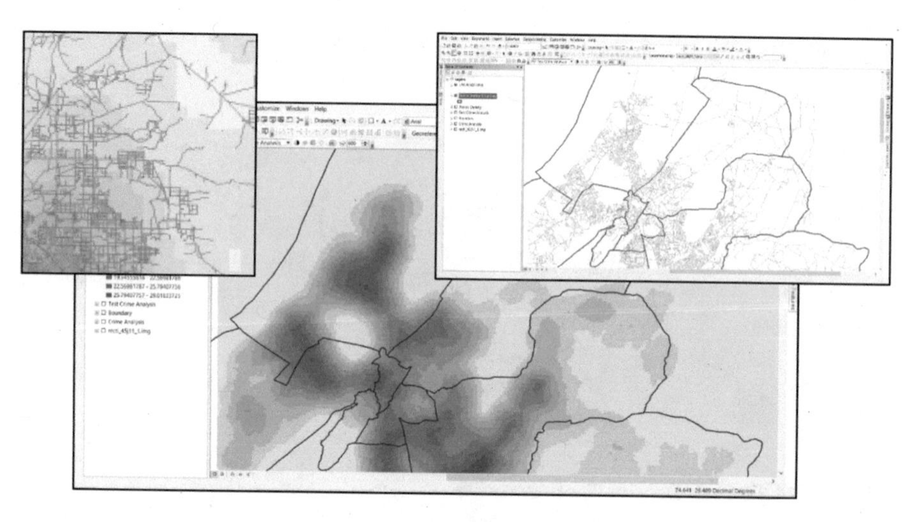

FIGURE 4.17 Application of Density Tool (Line Density).

(Wang, 2015). For hot spot analysis, it is very essential to have clustering within the data. The two methods mentioned will return values, including a Z-score, and when analyzed together will indicate if clustering is found in the data or not to help further processing of the data. Data will need to be aggregated to polygons or point of incident convergence before performing the spatial autocorrelation analysis as in case of IDW mentioned in the earlier section. Hot spot analysis, also known as Getis-Ord Gi* (G-I-star), works by looking at each feature in the data set within the context of neighboring features in the same data set. There may be a feature with a high value but it may not be a statistically significant hot spot. In order to be a significant hot spot, a feature with a high value will be surrounded by other features with high values (Geospatial Intelligence, 2019). "The local sum for a feature and its neighbors is compared proportionally to the sum of all features; when the local sum is very different from the expected local sum, and that difference is too large to be the result of random choice, a statistically significant z-score results." A Z-score and a P-value are returned for each feature in the data set.

A high Z-score and a low P-value for a feature indicate a significant hot spot in the area. A low negative Z-score and a small P-value indicate a significant cold spot (Fig. 4.18). The higher (or lower) the Z-score, the more intense the clustering. A Z-score near 0 means no spatial clustering. The output of the analysis tells us where features of either high or low values cluster spatially. Scale is important here, as there may be noticeable regional differences between the administrative boundaries and the grid (below) and the method one chooses would depend upon the data and scale of analysis.

The minimum number of features being analyzed should usually be 30 as results are unreliable sub-30. When using the grid approach, it is advisable to remove grid values of zero before performing the hot spot analysis.

Any statistical analysis requires that the data is in a projected coordinate system using a unit of measurement. A popular projection is the UTM Zone with the data

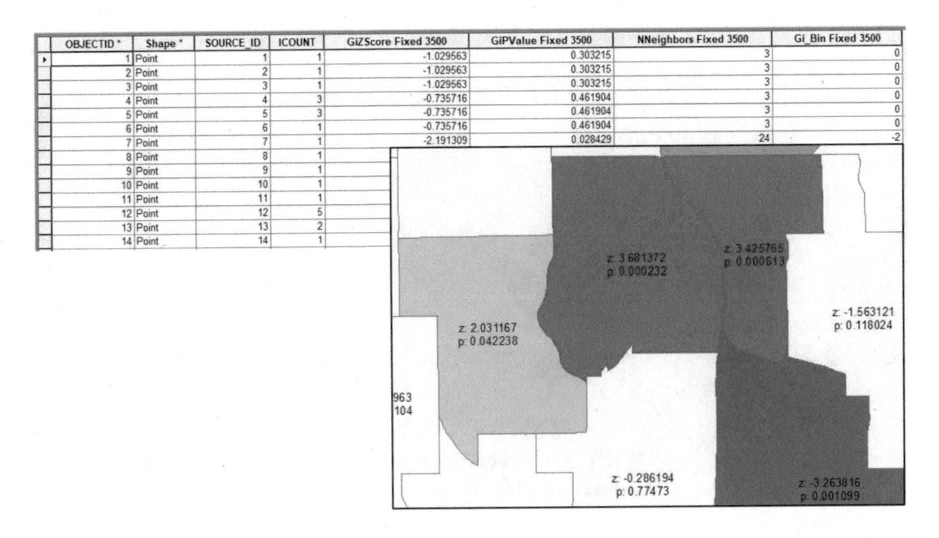

OBJECTID *	Shape *	SOURCE_ID	ICOUNT	GiZScore Fixed 3500	GiPValue Fixed 3500	NNeighbors Fixed 3500	Gi_Bin Fixed 3500
1	Point	1	1	-1.029563	0.303215	3	0
2	Point	2	1	-1.029563	0.303215	3	0
3	Point	3	1	-1.029563	0.303215	3	0
4	Point	4	3	-0.735716	0.461904	3	0
5	Point	5	3	-0.735716	0.461904	3	0
6	Point	6	1	-0.735716	0.461904	3	0
7	Point	7	1	-2.191309	0.028429	24	-2
8	Point	8	1				
9	Point	9	1				
10	Point	10	1				
11	Point	11	1				
12	Point	12	5				
13	Point	13	2				
14	Point	14	1				

FIGURE 4.18 Z-Score and P-Score Values in Hot Spot Analysis.

using meters (m) as the unit of measurement. The result of the hot spot analysis tool is a new feature class where every feature in the data set is symbolized based on whether it is part of a statistically significant hot spot, a statistically significant cold spot, or is not part of any statistically significant cluster. In crime analysis, a hot spot is an area that has a greater-than-average number of criminal or disorder events, or an area where people have a higher-than-average risk of victimization. An area can be considered a hot spot if a higher-than-average occurrence of the event being analyzed is found in a cluster, and cooler to cold spots with less-than-average occurrences. The higher above the average with similar surrounding areas, the hotter the hot spot.

The hot spot analysis tool assesses whether high or low values (the number of crimes, accident severity, or disease incidences, for example) cluster spatially or not. The field containing those values is the Analysis Field. For point incident data, however, we basically assess the incident intensity rather than analyze the spatial clustering of any particular value associated with the incidents. In that case, it is required to aggregate the incident data prior to analysis. There are several ways to do it. If there are polygon features for the study area, the spatial join tool is used to count the number of events in each polygon and the resultant field containing the number of events in each polygon becomes the input field for analysis. While performing the analyses, the create fishnet tool is used to construct a polygon grid over all point features. Later, the spatial join tool is used to count the number of events falling within each grid polygon. If any grid polygons are falling outside the study area, they can be removed. In cases where many of the grid polygons within the study area contain zeros for the number of events, increase the polygon grid size, if appropriate, or remove those zero-count grid polygons prior to analysis.

Similarly, if the data is in the form of number of coincident points or points within a short distance of one another, the Integrate with the Collect Events Tool is

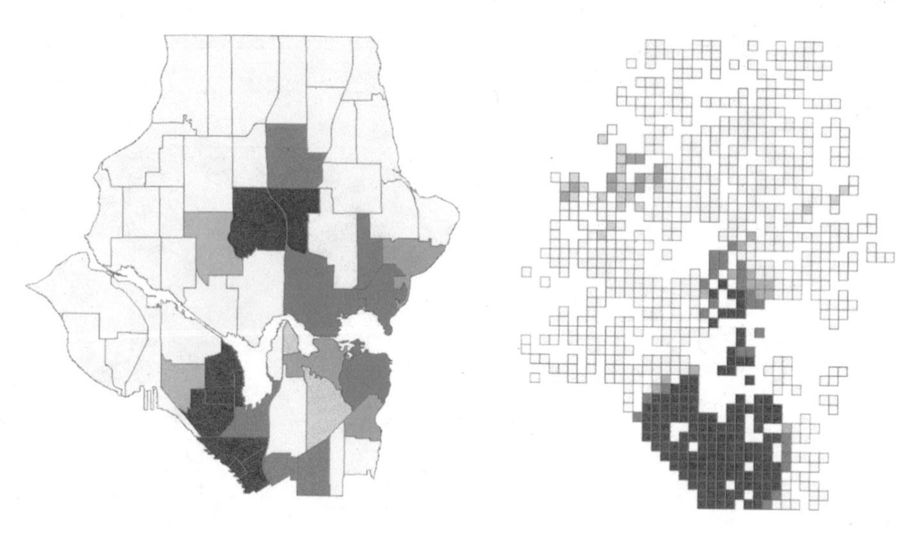

FIGURE 4.19 Grid Approach (Hot Spot Analysis).

used to (1) snap features within a specified distance of each other together, and later (2) create a new feature class containing a point at each unique location with an associated count attribute to indicate the number of events/snapped points. The resultant ICOUNT field as your Input Field for analysis is used.

In Fig. 4.19, the red areas are hot spots, or areas where high numbers of crime incidents are surrounded by other areas with high numbers of crime incidents. The blue areas are cold spots, or areas where low numbers of crime incidents are surrounded by other areas with low numbers of crime incidents. The beige shaded areas are not part of statistically significant clusters. The following figure (Fig. 4.20) discusses the application of hot spot analysis in crime mapping in Ajmer City, Rajasthan, India.

The hot spot concentration is confined in the north-central part of the city. It has high value concentration in the old city region with 99% confidence and has 95% and 90% confidence in the surrounding region. The results clearly earmark the crime-occurring region in the core city area with complex transportational network, narrowing down of streets, high population concentration, ferry markets possibly the high non-working population, illiterate people of daily wage jobs, and street/footpath vendors of low per-capita. The other supporting field observation is that intake of locally made liquor and drugs.

4.3 REGRESSION ANALYSIS

The regression analysis facilitates a researcher to model, examine, and explore spatial relationships, and can help explain the factors behind their observed spatial patterns (Huang, 2017). It is basically used for prediction; for example, to understand why a particular type of crime such as molestation is occurring in a certain region or to predict rainfall where there are no rain gauges.

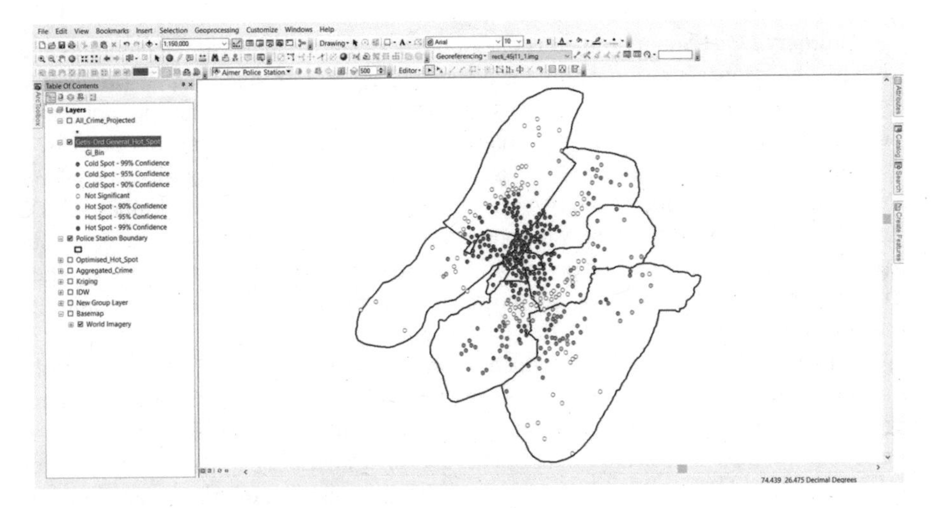

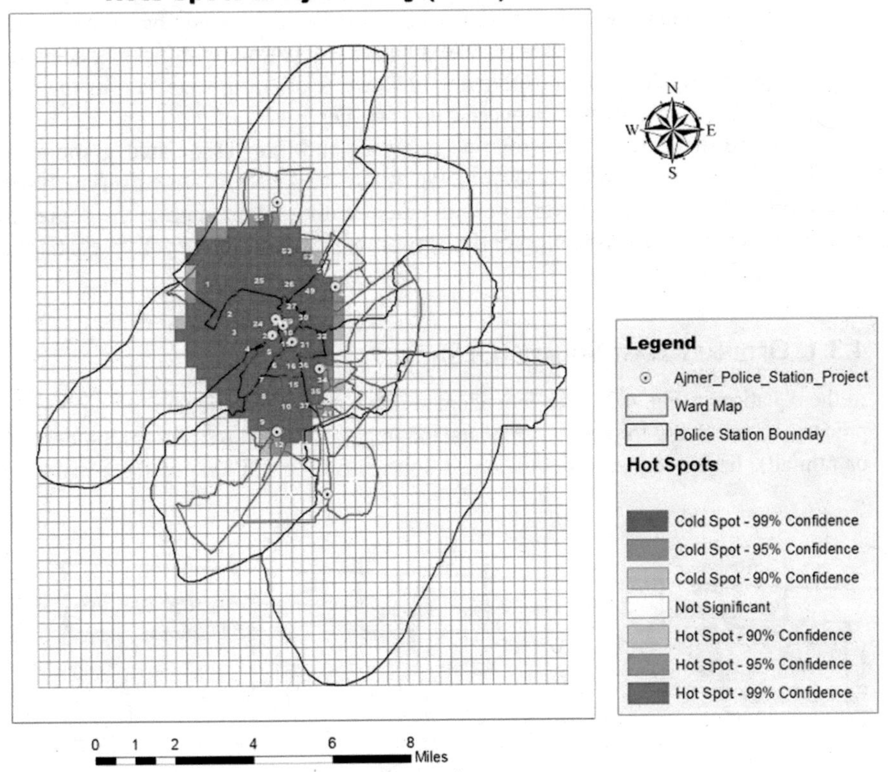

FIGURE 4.20 Crime Hot Spot Analysis in Ajmer City, India (2011).

Ordinary Least Square (OLS) is one of the best-known regression techniques and is also the proper starting point for all spatial regression analyses. It provides a global model of the variable or process we are trying to understand or predict (early death/rainfall) and creates a single regression equation to represent that process.

Geographically Weighted Regression (GWR) is one other popular spatial regression technique, increasingly used in geography and other disciplines. GWR provides a local model of the variable or process that we are trying to understand/predict by fitting a regression equation to every single feature of the data set. When used properly, these methods are powerful and reliable statistics for examining/estimating linear relationship.

It is observed that linear relationships are either positive or negative. If we observe that the number of search and rescue events increases when daytime temperatures rise, the relationship is said to be positive; there is a positive correlation. It can also be said that search and rescue events decrease as daytime temperatures decrease. In crime studies, if we observe that the number of crimes go down as the number of police officers patrolling an area goes up, the relationship is said to be negative. This negative relationship can also be expressed by stating that the number of crimes increases as the number of patrolling officers decreases. Fig. 4.21 below depicts both positive and negative relationships, as well as the case where there is no relationship between two variables.

The figure discusses the correlation analyses and their associated graphics to test the strength of the relationship between two variables. Regression analyses, on the other hand, make a stronger claim and attempt to demonstrate the degree to which one or more variables potentially promote positive or negative change in another variable.

4.3.1 Ordinary Least Square (OLS)

In the equation in Fig. 4.22, the dependent variable (y) is the variable representing the process we are trying to predict or understand (e.g., residential robbery, rapes, density, or rainfall). In the regression equation, it appears on the left side of the equal sign.

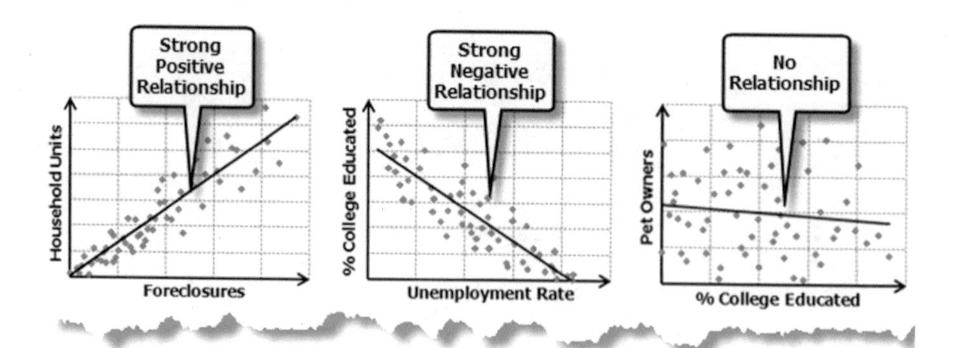

FIGURE 4.21 Basics of Correlation Analysis.

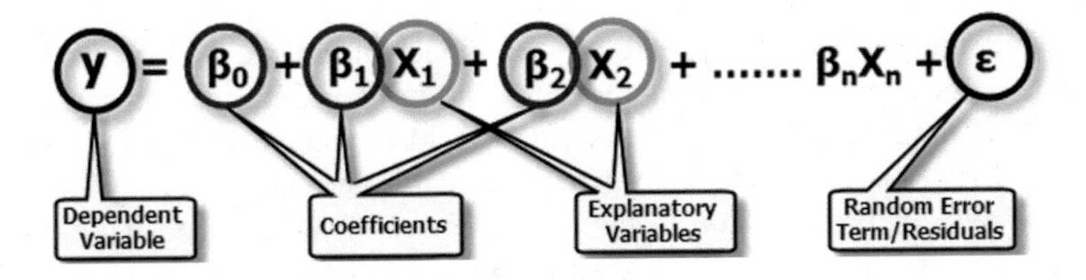

FIGURE 4.22 Regression Equation.

While you can use regression to predict the dependent variable, you always start with a set of known *y* values and use these to build (or to calibrate) the regression model. The known *y* values are often referred to as observed values.

Independent/explanatory variables (*x*) are the variables used to model or to predict the dependent variable values. In the regression equation, they appear on the right side of the equal sign and are often referred to as explanatory variables. The dependent variable are functions of the explanatory variables. To predict annual purchases for a proposed store, we might include the number of potential customers, distance to competition, store visibility, and local spending patterns, for example, as the explanatory variables in our model.

Regression coefficients (β) are computed by the regression tool. They are values, one for each explanatory variable, that represent the strength and type of relationship the explanatory variable has with the dependent variable. If we are finding out the fire frequency as a function of solar radiation, vegetation or precipitation, crime analysis etc. We tend to expect a positive relationship between fire frequency and solar radiation (the more sun, the more frequent the fire incidents). When the relationship is positive, the sign for the associated coefficient is also positive. We tend to expect a negative relationship between fire frequency and precipitation (places with more rain have fewer fires). Similarly coefficients for negative relationships have negative signs. When the relationship is a strong one, the coefficient is large and weak relationships are associated with coefficients near zero. β_0 is the regression intercept. It represents the expected value for the dependent variable if all of the independent variables are zero.

P-values are the probability computed by most regression methods, for the coefficients associated with each independent variable. The null hypothesis for this statistical test states that a coefficient is not significantly different from zero (in other words, for all intents and purposes, the coefficient is zero and the associated explanatory variable is not helping your model). Small P-values reflect small probabilities, and suggest that the coefficient is, indeed, important to your model with a value that is significantly different from zero (the coefficient is NOT zero). If we say that a coefficient with a P-value of 0.01, for example, is statistically significant at the 99 confidence level; the associated variable is an effective predictor. Variables with coefficients near zero do not help predict or model the dependent variable; they are almost removed from the regression equation, unless there are strong theoretical reasons to keep them.

R^2/R-Squared-Multiple R-Squared and Adjusted R-Squared are both statistics derived from the regression equation to quantify model performances. The values of

R-squared ranges from 0 to 100%. If our model fits in the observed dependent variable values perfectly, R-squared is 1.0 (and we no doubt, have made an error; perhaps we have used a form of *y* to predict *y*). More likely, we will see R-squared values such as 0.49, for example, which we can interpret by saying: this model explains 49% of the variation in the dependent variable. To understand what the R-squared value is getting at, create a bar graph showing both the estimated and observed *Y* values sorted by the estimated values. We notice and analyze how much overlap there is. This graphic provides a visual representation of how well the model's predicted values explain the variation in the observed dependent variable values. The Adjusted R-Squared value is always observed a bit lower than the Multiple R-Squared value because it reflects model complexity (the number of variables) as it relates to the data.

Residuals are the unexplained portion of the dependent variable, represented in the regression equation as the random error term, ε. The known values for the dependent variable are used to build and to calibrate the regression model. Using known values for the dependent variable (y) and known values for all of the explanatory variables (the xs), the regression tool constructs an equation that will help predict those known y values, as well as possible. The predicted values rarely match the observed values exactly. The differences between the observed y values and the predicted y values are called the residuals. The magnitude of the residuals from a regression equation is one measure of model fit where the large residuals indicate poor model fit.

To build a regression model is an iterative process that involves finding effective independent variables to explain the process we are trying to model/understand, then running the regression tool to determine which variables are effective predictors and then removing/adding variables until we find the best model possible.

The Exploratory Regression tool evaluates all possible combinations of the input candidate explanatory variables, looking for OLS models that best explain the dependent variable within the context of user-specified criteria (Dixon and Uddameri, 2016). One can access the results of this tool (including the optional report file) from the results window.

The primary output for this tool is a report file which is written to the results window. By right-clicking on the messages entry in the results window and selecting view will display the Exploratory Regression summary report in a message dialog box. This tool optionally creates a text file report summarizing the results. This report file is added to the Table Of Contents (TOC) and can be viewed in ArcMap by right-clicking on it and selecting open. This tool also produces an optional table of all models meeting our maximum coefficient P-value cut-off and Variance Inflation Factor (VIF) value criteria. The full explanation of the report elements and table is provided in Interpreting Exploratory Regression Results (Fig. 4.23). This tool uses both Ordinary Least Squares (OLS) and Spatial Autocorrelation (Global Moran's *I*). The optional spatial weights matrix file is used with the Spatial Autocorrelation (Global Moran's *I*) tool to assess model residuals but it is not used by the OLS tool. It tries every combination of the Candidate Explanatory Variables entered, looking for a properly specified OLS model. Only when it finds a model that meets our threshold criteria for Minimum Acceptable Adj

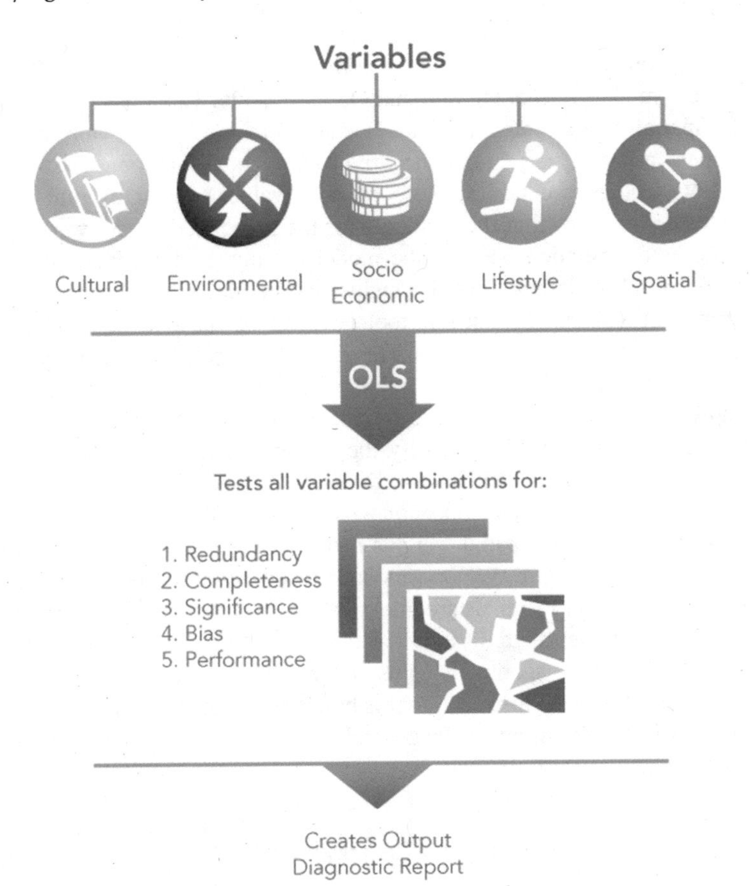

FIGURE 4.23 The Exploratory Regression Tool.

R-Squared, Maximum Coefficient P-value Cutoff, Maximum VIF Value Cutoff, and Minimum Acceptable Jarque-Bera P-value will it run the Spatial Autocorrelation (Global Moran's *I*) tool on the model residuals to see if the under/over-predictions are clustered or not. In order to provide at least some information about residual clustering in the case where none of the models pass all of these criteria, the Spatial Autocorrelation (Global Moran's *I*) test is also applied to the residuals for the three models that have the highest Adjusted R2 values and the three models that have the largest Jarque-Bera P-values.

When there is strong spatial structure in your dependent variable, we will want to try to come up with as many candidate spatial explanatory variables as you can. Some examples of spatial variables would be distance to major highways, accessibility to job opportunities, number of local shopping opportunities, connectivity measurements, or crime densities. Until we find explanatory variables that capture the spatial structure in our dependent variable, model residuals will likely not pass the spatial autocorrelation test. The significant clustering in regression residuals, as determined by the Spatial Autocorrelation (Global Moran's *I*) tool, indicates model misspecification. The Spatial Autocorrelation (Global Moran's *I*) is not run for all of the models tested, the optional Output Results Table has missing data for the SA (Spatial Autocorrelation) field. Because .dbf files do not store null values, these appear as very, very small (negative) numbers (something like -1.797693e +308). For geodatabase tables, these missing values appear as null values. A missing value indicates that the residuals for the associated model were not tested for spatial autocorrelation because the model did not pass all of the other model search criteria. The default spatial weights matrix file is used to run the Spatial Autocorrelation (Global Moran's *I*) tool is based on an 8 nearest neighbor conceptualization of spatial relationships. This default was selected primarily because it executes fairly quickly. To define neighbor relationships differently, however, we can simply and create our own spatial weights matrix file using the Generate Spatial Weights Matrix File tool, then specify the name of that file for the Input Spatial Weights Matrix File parameter. Inverse Distance, Polygon Contiguity, or K Nearest Neighbors, are all appropriate conceptualizations of spatial relationships for testing regression residuals. The spatial weights matrix file is only used to test model residuals for spatial structure. When a model is properly specified, the residuals are spatially random (large residuals are intermixed with small residuals; large residuals do not cluster together spatially). When there are 8 or less features in the Input Features, the default spatial weights matrix file used to run the Spatial Autocorrelation (Global Moran's *I*) tool is based on K nearest neighbors where K is the number of features minus 2. In general, you will want to have a minimum of 30 features when you use this tool.

4.3.2 Geographically Weighted Regression (GWR)

The Geographically Weighted Regression (GWR), a local form of linear regression is used to model spatially varying relationships.

It constructs a separate equation for every feature in the data set incorporating the dependent and explanatory variables of features falling within the bandwidth of each target feature. The shape and extent of the bandwidth is dependent on the user input for Kernel type, Bandwidth method, Distance, and Number of Neighbors parameters with one restriction. When the number of neighboring features would exceed 1,000, only the closest 1,000 are incorporated into each local equation. GWR is usually be applied to data sets with several hundred features for best results and is not an appropriate method for small data sets. It does not work with multipoint data. The GWR tool produces a variety of different outputs. Right-clicking on the Messages entry in the Results window and selecting view will display a GWR tool execution summary report. The GWR tool also produces an Output feature class and a table with the tool execution summary report diagnostic values. The name of this table is automatically generated using the output feature class name and its suffix. The Output feature class is also automatically added to the table of contents with a hot/cold rendering scheme applied to model residuals and a full explanation of each output is provided in interpreting GWR results (Fig. 4.24). In global regression models, such as OLS, results are unreliable when two or more variables

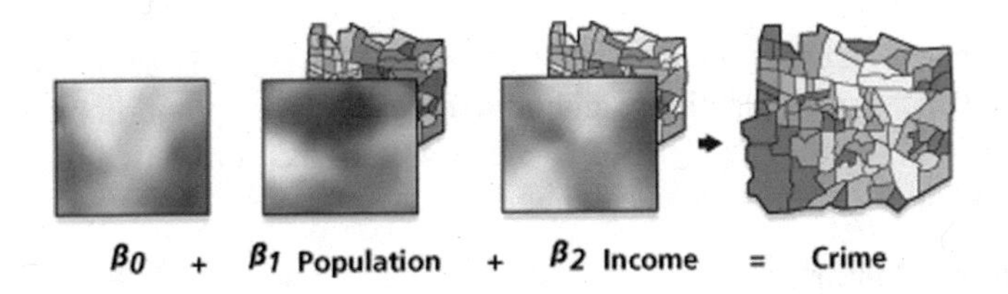

FIGURE 4.24 Geographically Weighted Regression.

exhibit multicollinearity (Stewart et al., 2002). GWR builds a local regression equation for each feature in the available data set. When the values for a particular explanatory variable cluster spatially, we will very likely have problems with local multicollinearity. In the earlier case, the output feature class indicates when results are unstable due to local multicollinearity. So one should not trust results for features with a condition number larger than 30, equal to Null or, for shapefiles, equal to -1.7976931348623158e+308.

4.3.3 Applications of Regression Analysis

This tool is effectively used for modeling fire frequency to determine high-risk areas and to understand the factors and their degree of impact that contribute to high-risk areas. Analyzing property loss from fire as a function of variables such as degree of fire department involvement, response time, property value, etc. Crime incidences can be investigated by taking population density, occupational structure, poverty index, etc., as the related variables. Regression analysis basically aims to measure the extent that changes in one or more variables jointly affect changes in another. Example: Understand the key characteristics of the habitat for some particular endangered species of birds (perhaps precipitation, food sources, vegetation, predators) to assist in designing legislation aimed at protecting that species in a particular region.

This tool helps to model some phenomena in order to predict values for that phenomenon at other places or other times. It aims to build a prediction model that is consistent and accurate. Example: When are real estate values likely to go up in the next coming year? Another example could be, there are rain gauges at particular places and a set of variables that explain the observed precipitation values. Now to understand and find out how much rain falls in places where there are no gauges, we can apply the regression tool. Regression may be used in cases where interpolation is not effective because of insufficient sampling; there are no gauges on peaks or in valleys, for example.

Regression analysis can also be used to test hypotheses. If we are modeling residential crime in order to better understand it, and hopefully recommend policy making and government interventions to prevent it, we have questions or hypotheses to test. For example, the broken window theory indicates that defacement of public property (graffiti, damaged structures, etc.) invite other crimes. So can we investigate a positive relationship between vandalism incidents and residential burglary? Is there a relationship between illegal drug use and robbery (might drug addicts steal to support their habits)? There can be analysis about occurrence of more incidences in residential neighborhoods with higher proportions of elderly or female-headed households. Or people are at greater risk for house break incidences if they live in a rich or a poor neighborhood.

Concluding Remarks

Crime mapping and analysis deals with identifying crime hot spots which are strategically very important locations for police surveillance and functioning as they enable a channelized focus on mobilization and resource use. Use of geospatial techniques such as spatial analysis help in the usage of spatial information to extract new information and meaning from the original data and is immensely applicable in research for terrain analysis, slope profiling, hydrological modeling, climatic studies, wildlife management, tourism sector, and many related fields. Interpolation techniques such as IDW and kriging estimates surface from a scattered set of points beneficial in drawing susceptible crime spots most appropriate. The Density tool also proves to be useful to create density surfaces to represent the distribution of a crime incidents from a set of observations, or the degree of urbanization of an area based on the density of roads.The hot spot analysis tool assesses whether high or low values (the number of crimes, accident severity, or disease incidences, for example) cluster spatially or not, giving a visual impressions for crime investigations. The Regression analysis facilitates a researcher to model, examine, and explore spatial relationships, and helps crime analysts to explain the factors behind their observed spatial patterns. Many crime studies investigate crime incidences taking population density, occupational structure, poverty index, etc., as the related variables.

REFERENCES

Bhattacharjee, S. (2019) *Semantic Kriging for Spatio-temporal Prediction.* Springer Nature, Singapore.

Dixon, B. and Uddameri, V. (2016) *GIS and Geocomputation for Water Resource Science and Engineering*, 1st ed. American Geophysical Union, Oxford.

Geospatial Intelligence. (2019) *Concepts, Methodologies, Tools, and Applications*, edited by Management Association, Information Resources, IGI Global, USA.

Huang, B. (2017) *Comprehensive Geographic Information Systems.* Elsevier.

Mohamed, A. and Nasef, M. R. (2019) *A Geographic Information System Approach for Mapping and Assessing the Climate Change.* Science and Arts Publishers Inc, New Delhi.

Rocha, A., Adeli, H., Paulo, L., and Costanzo, S. (2018) *Trends and Advances in Information Systems and Technologies.* Springer, New York, NY, Vol. 1, 2018.

Shekhar S. and Hui X. (2008) *Encyclopedia of GIS.* Springer Science & Business Media, New York, NY.

Stewart, A., Fotheringham, Brunsdon, C., and Charlton, M. (2002) *Geographically Weighted Regression: The Analysis of Spatially Varying Relationships.* John Wiley & Sons, New York.

The ESRI. (2005) *Guide to GIS Analysis Vol. 2: Spatial Measurements & Statistics.* Andy Mitchell, ESRI Press.

Wang, F. (2015) *Quantitative Methods and Socio-Economic Applications in GIS.* CRC Press, Taylor and Francis Group, Boca Raton, FL.

5 Crime Mapping and Geospatial Analysis

The advancement of geospatial technologies has made GIS a trusted partner for policing and planning to curb crime in society. This chapter discusses the relationship between crime with demographic parameters and discusses the approaches of crime analysts of overlaying other data sets of transportational models to better understand the underlying causes of crime. Network analysis in criminal intelligence is used to organize data and reveal patterns in the nature and extent of relationships between data points and help law enforcement administrators to devise strategies to deal with the problem. The network analysis uses a set of approaches to understand and act against serious crime occurrences, criminal groups, and criminal markets. These approaches are based on link analysis and increasingly includes techniques from social network analysis. Individual demographic characteristics and aggregate population processes are central to many theoretical perspectives and empirical models of criminal behavior. Researches underscores the importance of criminal and deviant behavior for understanding the demography of the life course and macrolevel population processes. The chapter focuses on how demography affects crime, discussing the various parameters of intervention such as impact of age, sex, and race on criminal behavior, and the difference between compositional and contextual effects of demographic structure on aggregate crime rates. The intersection of criminal and demographic events in the life course, and the influence of criminal victimization and aggregate crime rates on residential mobility, migration, and population redistribution are some pertinent issues of analysis.

5.1 DEMOGRAPHIC PATTERNS AND PROFILES

Demographic study provides an insight and directions for future research on many linkages between criminal and demographic behavior (Scott J. South and Steven F. Messner, 2000). The word "demographics" originated in the ancient Greek civilization, where it was used to refer to *demo* for people and *graphics* for measurement. Demography is fundamentally a numerical and statistical study addressing the population considering human beings in focus. It studies demography as a science of dynamic living population that changes over time or space. Demographics are features quantifiable for a given population. Demographic thoughts dated back to ancient times and were present in many cultures and civilizations, such as ancient Greece, ancient Rome, China, and India. This can be found in ancient Greece in works by Herodotus, Thucydides, Hippocrates, Epicurus, Protagoras, Polus, Plato, and Aristotle. Formal demography restricts its objective of study to the measurement of population processes, while the wider field of social dynamics or population studies often analyzes the relationships that affect a population between economic, financial, cultural, and biological processes. In a broader sense, the branch of science dealing with population is demography and the people who study this science are referred as demographers; they intend to study the dynamics of the population by inquiring three major processes, the first one is birth, the second is migration, and the third is aging, which includes death. The contribution given by the above three processes go to population changes which includes inhabitation, building nations and societies, and developing different cultures. In a majority of the countries, birth rates below the replacement level are 2.1 children/women along with growing life expectancy which as overall described with term "the aging of societies." Demographic studies should not be misunderstood as a tool to assist politics in terms of change in demography; instead it is study of the factors, causes, and phenomenon which are related to the demographic change and social dynamism.

5.1.1 Demography

Demography is defined as the study of a population or populations, particularly with reference to size and density, fertility, mortality, growth, age distribution, migration, vital statistics, and the interaction of all of these with social and environmental conditions. According to Barckley, "The numerical portrayal of human population is known as demography." Similarly, according to Thomson and Lewis, "The population student is interested in population's size, composition and distribution; and in changes in these aspects through time and causes of these changes. Demography constructs the simplest processes imaginable" (Hervé, 2008).

The above definitions take a constricted approach as addressing only the quantitative facet of demography. There are some other writers who took a different approach addressing both the quantitative and qualitative facets pertaining to population studies. In this context, according to Hauser and Duncan, "Demography is the study of size, territorial distribution and composition of population, changes therein, and the components of such changes, which may be identified as fertility, mortality, territorial movement (migration), and social mobility (change of status)" (Gordon, 2016).

According to Frank Lorimer, "In broad sense, demography includes both demographic analysis and population studies. A broad study of demography studies both qualitative and quantitative aspects of population" (Pathak, 1998).

Thus, according to Donald J. Bougue, "Demography is a statistical and mathematical study of the size, composition, spatial distribution of human population, and of changes overtime in these aspects through the operation of the five processes of fertility, mortality, marriage, migration and social mobility (Donald, 1969). Although it maintains a continuous descriptive and comparative analysis of trends, in each of these processes and in its net result, its long run goal is to develop a body of theory to explain the events that it charts and compares." The above definitions are panoptic as they consider human migration and impact of the education, employment, social status, etc., on the change in the status of population along with size, composition, and distribution of population.

5.1.2 Characteristics of Demography

5.1.2.1 Size and Shape of Population

The term "population" in its simplest manner can be defined as the number of people in a city or town or area, region, country, or entire world. While denoting the size of population of any area or region, the total number of persons residing in a definite area at a definite time are considered. Factors that may vary are size and shape of population of any region, state, or nation. The reason behind the foresaid changing factors is every country has its own unique customs, specialities, socioeconomic conditions, cultural atmosphere, moral values, and different standards for acceptance of artificial means of family planning and availability of health facilities, etc.

5.1.2.2 Aspects Related to Birth Rate and Death Rate

While studying the population two factors play decisive role, they are birth rate and death rate, as they immensely influence the size and shape of the population. The birth rate and death rate are further affected by marriage rate, belief regarding social status and marriage, age of marriage, orthodox customs related to marriage, early marriage and its effects on the health of the mother and the child, child infanticide rate, maternal death, still birth, resistance power, level of medical services, availability of nutritious food, purchasing power of the people, etc.

5.1.2.3 Composition and Density of Population

The study of composition and density of population is prominent in the subject matter of demography. The other related factors are sex ratio, race wise, and age-group wise size of population, the ratio of rural and urban population, distribution of population according to religion and language, occupational distribution of population, agricultural and industrial structure, and per sq. km. density of population.

5.1.2.4 Socioeconomic Problems

One of the major problems related to population growth is the effects of high density in urban areas due to industrialization are of greater importance because they affect people's socioeconomic lives. Problems such as slum areas, toxic air and water, violence, drug abuse, juvenile delinquency and prostitution, are also important demographic study topics.

5.1.3 Quantitative and Qualitative Aspects

The qualitative aspects also play an important role in population studies along with quantitative problems. In addition, demographic studies also include number of hospitals, total number of beds, availability of doctors in the population, life expectancy at birth, daily availability of minimum calories, resistance and immunity power, the changes brought in the attitudes of people regarding child birth, and availability of proper medical facility for delivery, etc., as per the investigative question. Demographic factors have been cited as the strongest determinants of crime rates and hence have been central to crime predictions.

The key demographic variable appears to be the size of the male population within the crime-prone years of, for example, between 15 to 25 years of age. In those societies with large proportions of young males, there tends to be a higher crime rate. A number of studies have shown that crime rates are also closely tied to the strength of the economy; during economic recessions, property crime rates are inclined to grow rapidly, whereas during more economically favorable periods, they have a tendency to fall. Demography is considered as a science on all counts. Not only is it a positive science of "what is" but it is also a normative science of "what should be." It studies the causes and effects of population problems and also suggests policy measures to resolve them. To conclude with Irene Teubner, "With improved data, new techniques and precise measurement of the demographic transition that is occurring, demography has become a science. In fact, it has become an applied science and applied technology."

5.2 DEMOGRAPHIC DIVIDEND OR A DEMON

According to the census reports of Indian Census 2011, the population of India is 1,210,854,977 with 623,724,248 males and 586,469,174 females. During 2020, the India population is projected to increase by 17,466,075 people and reach 1,404,763,527 in the beginning of 2021. Fig. 5.1 indicates the state-wise population distribution in the country. The total literacy rate in the country at present is 74.04%. The density of population is 382 persons/sq.km. In regards to sex ratio, at present there are 940 females per 1,000 males and the child sex ratio is 914 females per 1,000 males.

As of 1 January 2020, the population of India was estimated to be 1,387,297,452 people. This is an increase of 1.26% (17,248,911 people) compared to population of 1,370,048,541 the year before. In 2019, the natural increase was positive, as the number of births exceeded the number of deaths by 17,810,631. Due to external migration, the population declined by 561,720. The sex ratio of the total population was 1,070 (1,070 males per 1,000 females) which is higher than the global sex ratio. The global sex ratio in the world was approximately 1,016 males to 1,000 females as of 2019.

Below are the key figures for the Indian population in 2019:

- 27,962,691 live births
- 10,152,060 deaths
- Natural increase: 17,810,631 people
- Net migration: 561,720 people
- 717,105,446 males as of 31 December 2019
- 670,192,006 females as of 31 December 2019

It is essential to mention that the changing demographic profile of India has laid a measurable impact on the crime occurrence and increase in the country.

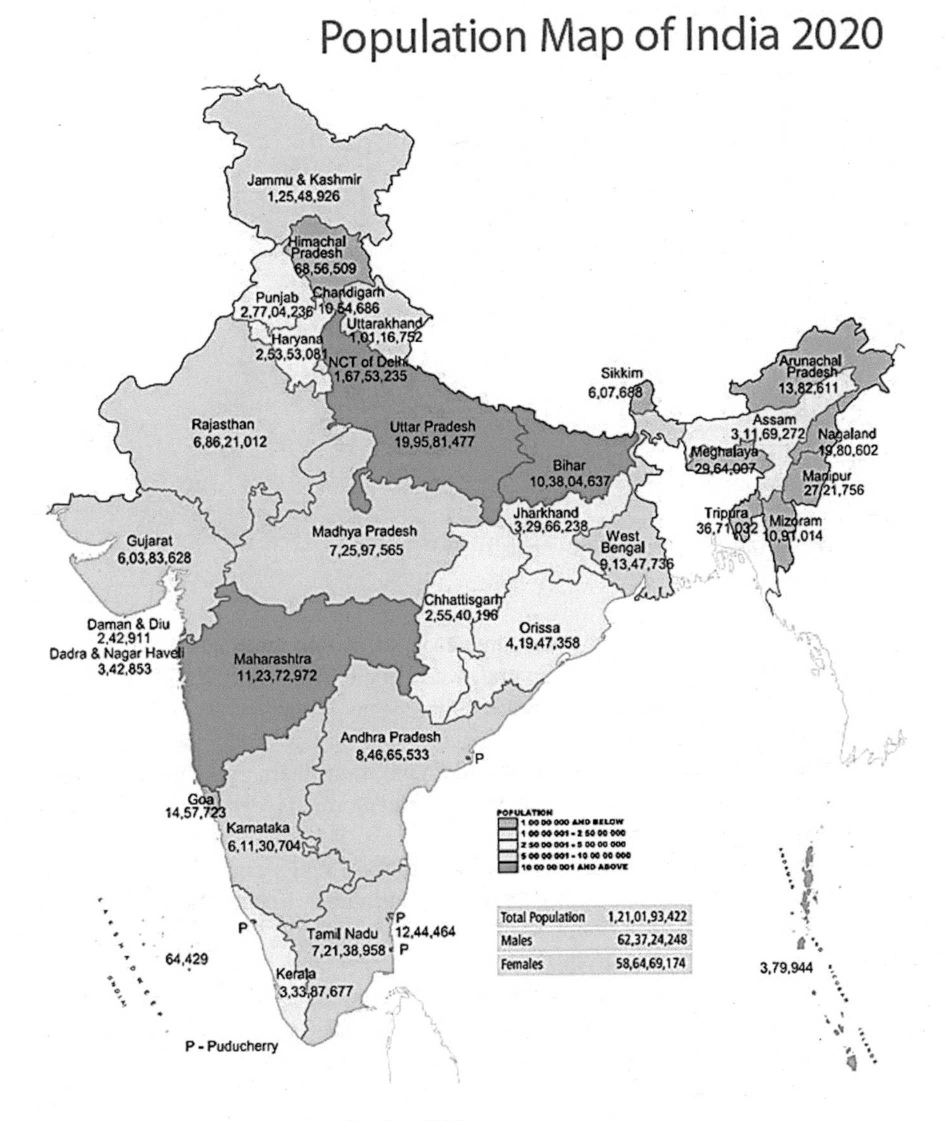

Total Population	1,21,01,93,422
Males	62,37,24,248
Females	58,64,69,174

FIGURE 5.1 Population Map of India, 2020.

Fig. 5.2 highlights the ranks of India's most populous cities in the world as per the WHO, Annual Average 2016. Recently, the NCRB has provided crime data in two sets: one on the states of India and one on the 19 largest "metropolitan cities" exceeding 2 million in population. It would be appropriate to point out that while the NCRB provides statistics on urban agglomerations with populations above 2 million, some of them stand high on the demographic ladder and go right up to populations of 20 million and beyond. The five largest are Delhi, Mumbai, Kolkata, Chennai, and Bengaluru.

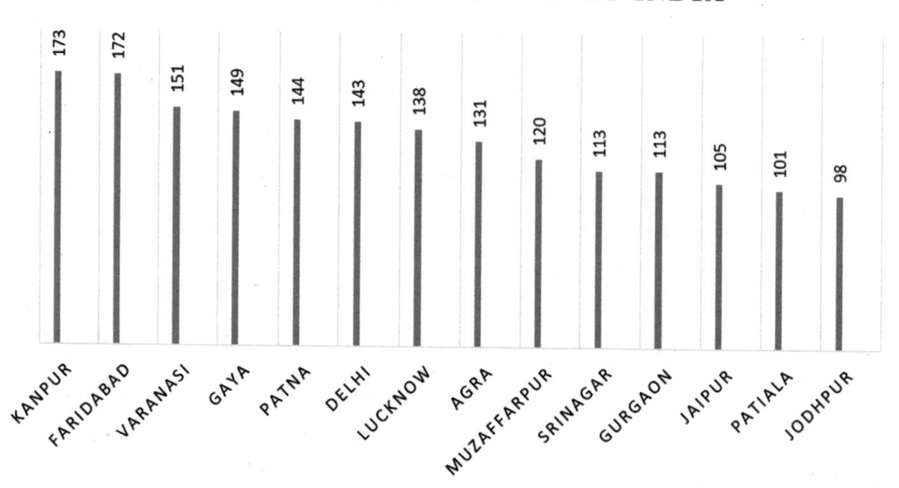

FIGURE 5.2 Most Populous Cities of India (WHO, 2016).

Of the 19 urban agglomerations, three each fall within the states of Maharashtra and Uttar Pradesh; two each in Gujarat, Kerala, and Tamil Nadu; and one each in Bihar, Delhi, Karnataka, Madhya Pradesh, Rajasthan, Telangana, and West Bengal. For comparison, we have clubbed together six urban agglomerations (Delhi, Ghaziabad, Jaipur, Kanpur, Lucknow, and Patna) as northern urban agglomerations and six others (Bengaluru, Chennai, Coimbatore, Hyderabad, Kochi, and Kozhikode) as southern urban agglomerations.

In terms of the Indian Penal Code (IPC) crimes, Delhi's rate of 1,306 (per lakh population) far exceeds any other urban agglomeration. Kochi, Patna, Jaipur, and Lucknow follow with crime rates of 809, 751, 683, and 600. The lowest crime rates with regard to IPC crimes are in Kolkata (141), Coimbatore (144), Hyderabad (187), Mumbai (212), and Chennai (221) (Fig. 5.3). The northern urban agglomerations in terms of IPC crime rates are about two times higher than southern urban agglomerations. In terms of fatal attacks that have resulted in deaths, Patna tops the list, with a crime rate of nine murders per lakh population. It is followed by Nagpur (8), and Indore, Jaipur, and Bengaluru (3 each). The lowest crime rates for murder are reported from Kozhikode, Kochi, Kolkata, Mumbai, and Hyderabad all clocking below one murder per lakh population. It is shocking to note that here the northern urban agglomerations have almost three times higher crime rates than the southern urban agglomerations.

As per Fig. 5.4, Delhi tops the crime rate chart for kidnappings and abductions with a rate of 32, followed by Indore (31), Patna (27), Lucknow (24), and Ghaziabad (23). The lowest crime rates are recorded by Coimbatore, Chennai, Kozhikode, Kochi, and Kolkata, all below three. Here, the northern block has a crime rate seven times higher than the south. In regards to crimes against women, Lucknow has the highest rate of 179, followed by Delhi (152), Indore (130), Jaipur (128), and Kanpur (118). The lowest rates are reported from Coimbatore (7), Chennai (15), Surat (28), Kolkata (29), and

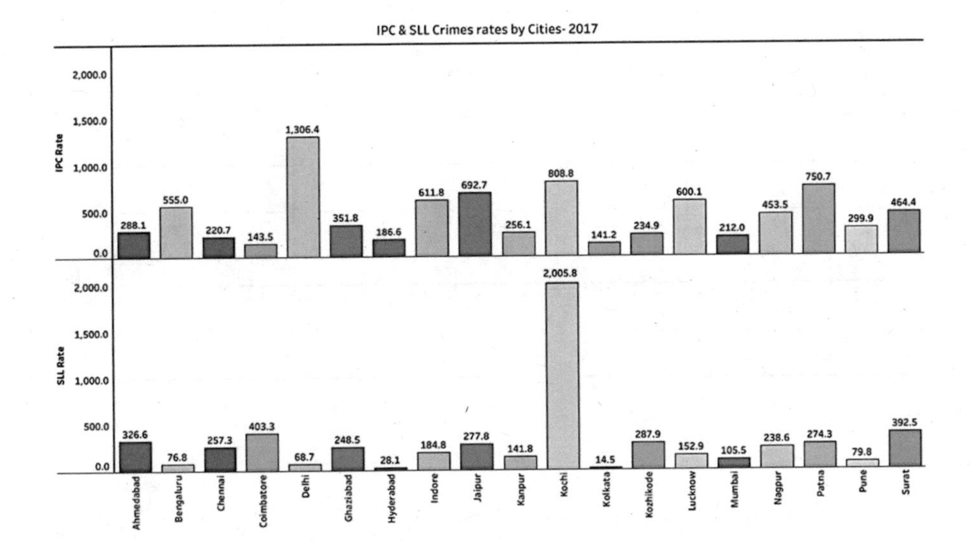

FIGURE 5.3 Crime Rates in Major Cities in India, 2017.

Kozhikode (33). Here again, the northern urban agglomerations have recorded a crime rate three times higher than the southern urban agglomerations.

For crimes against children, Delhi ranks the first with a crime rate of 35, followed by Mumbai (19), Bengaluru (8), Pune (7), and Indore (4). The lowest rates are recorded by Coimbatore, Ghaziabad, Patna, Kochi, and Kozhikode, all below one. Senior citizens are the worst off in Mumbai with a crime rate of 30, followed by Delhi (20), Ahmedabad (14), Chennai (13), and Bengaluru (5). The best urban agglomerations for the old are Kanpur, Indore, Kozhikode, Ghaziabad, and Patna, all below one. Here, as an exception, the northern block emerges kinder to senior citizens than the southern.

Jaipur in Rajasthan emerges as the leading urban agglomerations in terms of economic offenses (141), followed by Lucknow (65), Bengaluru (41), Delhi (30), and Kanpur (26). Coimbatore, Chennai, Kozhikode, Patna, and Ahmedabad rank the lowest in economic offenses, all below ten. A factor of more than two separates the northern urban agglomerations from the less crime-prone southern urban agglomerations. As the information technology hub of India, Bengaluru, leads in cyber crimes (32), followed by Jaipur (22), Lucknow (21), Kanpur (8), and Mumbai (7).

Chennai, Kozhikode, Coimbatore, Delhi, and Kolkata are at the bottom of the list with crime rates lower than two. An analysis of all the above data reveals that Delhi, Jaipur, Lucknow, Indore, and Patna have the highest average crime rates across crime categories among top urban agglomerations of the country. At the other end are Kozhikode, Coimbatore, Chennai, Kolkata, and Kochi with the lowest average crime rates among top urban agglomerations. It is generally acknowledged that cities have a greater propensity to crime and that megacities have a higher crime rate than smaller cities.

If the total crimes are taken into account, which includes those charged under the IPC and SLL (Special and Local Laws), New Delhi (Capital City) tops the list of IPC

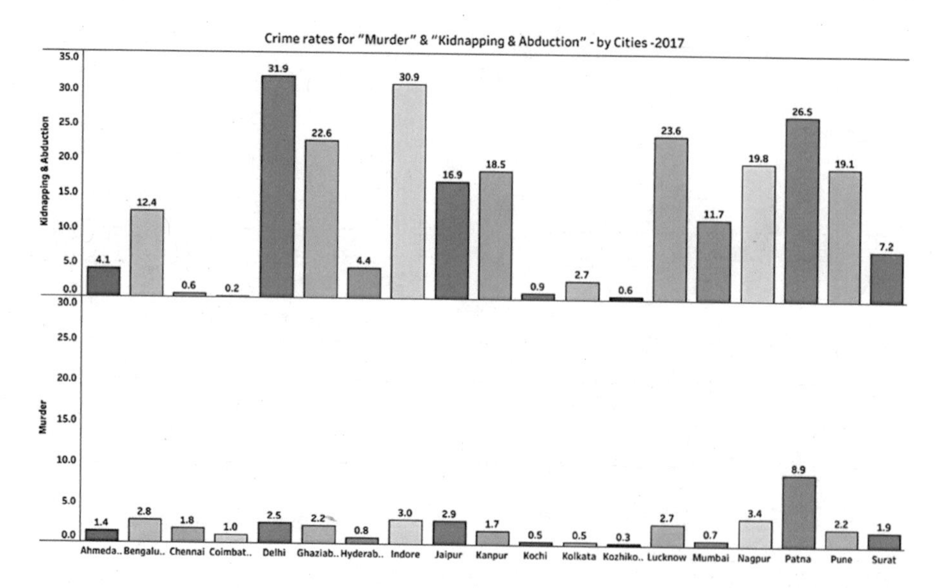

FIGURE 5.4 Crime Rate for Murder and Kidnapping and Abduction in Major Cities in India, 2017.

charged cases followed by Kochi and Patna whereas Kochi tops in Special and Local Laws (SLL) charged cases followed by Coimbatore and Surat (Fig. 5.3). New Delhi is way behind from Patna, Nagpur, Indore when it comes to murder rates. As per 2017 data on cognizable crime in major cities in India, Kochi leads with 2814 cases per 1 lakh population followed by Delhi (1375), Patna (1024) and Jaipur (970) (Fig. 5.5).

The drastic comparison between the demographic data and crime statistics reveal, that over 30% of youth between 15 and 29 years of age are idle, that is not in employment, education, or training providing them motivation and space for crime initiations. What, however, adds to this fuming churn is the 2011 Census finding that highlights 84 million literates and 33 million illiterates who are unemployed, and diploma holders are the highest among them. What is shocking then is the vagabond youth, with a piece of paper in their hands, that can't promise them a job. Team Lease reports that only 18% of those vocationally trained find work, but only 7% of these in formal employment. If small towns, rather than metros (with the exception of New Delhi) are where criminal activities are high, one needs to also factor in the kind of urbanization India is going through. Under equal conditions, a non-tier 1 city applicant has a 24% lower chance of finding work and can expect a salary that is Rs 66,000 less per annum.

The National Sample Survey also shows that the chances of getting a salaried job, that is one with some security, are much higher in larger cities than in smaller ones. But as metropolises in India, such as New Delhi and Mumbai, are showing a declining growth rate and Kolkata and Chennai may well have become stagnant, good jobs are getting harder to land in these places. Under such conditions, most of the job opportunities lie with the big-city networks who have access to these better jobs, and those who don't must, of necessity, go to smaller towns. In these smaller urban sites, the situation is different; jobs are not only more difficult to find they are

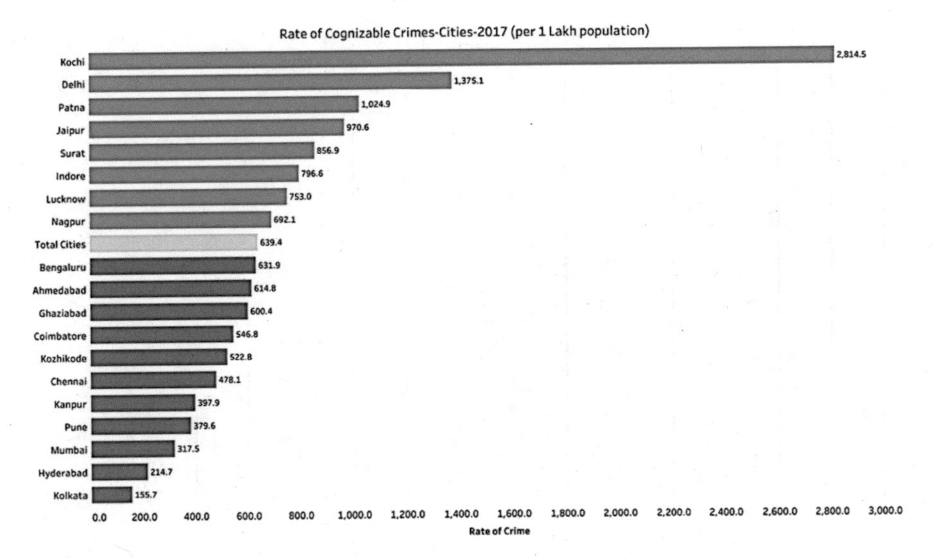

FIGURE 5.5 Rate of Recognizable Crime in Major Cities in India, 2017 (per 1 Lakh Population).

mostly informal and ill-paid ones. Consequently, towns such as Faridabad, Ghaziabad, Meerut, Patna, Pune, Jaipur, Kochi, Nagpur, Indore, and so on are not just getting bigger, but sickening too. This is because it is the literate unemployed who are largely circling this territory, searching for work and they have dreams much grander than what the unschooled can imagine. Unmet aspirations are socially more troublesome than empty stomachs.

Imagine what the atmosphere must be like when 15,000 graduates, in 2016, applied for the position of sweepers in Almroha, UP and when even this scaled down ambition fails, crime could easily become the default option. It is sad enough but true to mention, that what we are facing is not a demographic dividend but a "demographic demon." It is not a coincidence that juvenile criminals, the only category on which the crime bureau provides socioeconomic information, are overwhelmingly kids who have been to high school. Literate people migrate much more than illiterates do. Statistics reveal that the migration rate among those at the lowest income bracket is much below those who are better placed. In rural India, only 3% of those in the bottom income decile migrate, compared to 17% of those in the top. When it comes to urban male migration alone, the rate at which college graduates enter cities is more than double that of the "illiterates." If we try to connect the dots, we understand that the educated migrants move to small towns where well-paid jobs are rare and this leads to frustration, unsatisfactory behavior, and social anger which are fertile grounds for crime birth. Many researches have unveiled some astonishing facts such as places with high literacy rates, high human development index, and better standard of living have been registered for high crime rates in comparison to the rural setup in India.

Demographic characteristics and criminal behavior are intertwined in multiple ways at both the micro and macro levels. Another population characteristic that is

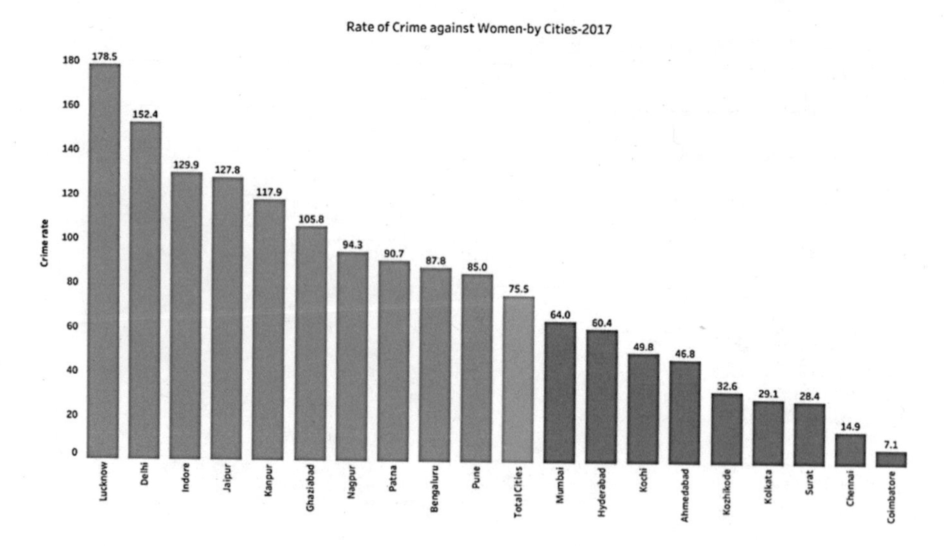

FIGURE 5.6 Rate of Crime Against Woman in Major Cities of India, 2017.

likely linked to aggregate crime rates is the sex ratio. In India, which has experienced a long-standing and likely worsening numerical deficit of girl counts, exploring the association between population sex composition and criminal victimization takes on special urgency. Prior studies of the impact of imbalanced sex ratios on crime, both in India and elsewhere, has generated equivocal findings.

Household level indicators of family structure, socioeconomic status, and caste as well as real indicators of women's empowerment and collective efficacy also emerge as significant predictors of self-reported criminal victimization and the perceived harassment of young women (Scott et al., 2014).

Although projecting trends in the above figure show crime against women indicate that criminal behavior is a risky undertaking, in India victimization rates are likely to increase along with the looming masculinization of the adult sex ratio. Researches indicate the observed associations between the sex ratio and criminal victimization risks are fairly modest; projected changes in the adult sex ratio are apt to increase annual odds of victimization risks by about 8% to 12% depending on the type of crime. At the same time, however, these seemingly small increases in *annual* risks may translate into at least moderate *lifetime* increases in victimization risk. With regard to the victimization of women, we find no support for the hypothesis, derived from Guttentag and Secord (1983), that men are less likely to harass women when women are relatively few in number. Rather, as with the other offenses, a numerical abundance of men and an attendant shortage of women tends to increase women's risk of being harassed in public spaces, often referred to as eve-teasing. Of course, the gendered dynamics described by Guttentag and Secord (1983) could conceivably temper the effects of offender supply, but even if so, these dynamics appear insufficiently powerful or extensive to override the aggregate criminogenic risks created by a numerical surplus of males.

The national data on crime as released by the National Crime Records Bureau (NCRB) indicates the crimes increase from 3,793 per million in 2016 to 3,886 per million in 2017, which means 100 more crimes took place per million people in 2017 compared to the previous year in India, though crimes such as murders and rapes have come down. (*The NCRB comes under the Union Ministry of Home Affairs and is responsible for collecting and analyzing crime data as defined by the Indian Penal Code (IPC) and special and local laws in India*).

A few states, in particular, Karnataka, Kerala, Tamil Nadu, Punjab, Rajasthan, and West Bengal showcased a decrease in crime rates, but most others continue to see a rise in crime occurrences. Incidents of theft have increased at the fastest rate. In Delhi, the crime rate rose by 8% in a year and is the fastest rate of growth among all states in India and translates to 11,500 crimes reported per million. Nearly 3 million Indian Penal Code (IPC) crimes and 2 million crimes under state laws were recorded in 2017. But even this report says, it is an understatement. *"The actual count of each crime per head may be underreported. This is because, among many offenses registered in a single FIR, only the most heinous crime (maximum punishment) will be considered as a counting unit,"* the report notes.

The report says incidence of rioting reduced from 53 crimes per million people in 2014 to 46 per million people in 2017. Incidence of kidnapping and abduction, on the other hand, rose from 62 per million to 74 per million. Crime rate under state laws that pertain mostly to prohibition, narcotics, excise, electricity-related ones, and gambling rose faster than crimes under the Indian Penal Code (IPC). IPC crime incidence also rose, with crimes such as kidnapping and attempt to murder on the rise, per million population. While the incidence of rapes has reduced, the overall crime rates against women continues to rise. Cases of rapes reduced from 63 per million people in 2016, to 52 per million in 2017, but a total of 3.5 lakh cases of crime against women were registered; these include murder, rape, dowry death, suicide abetment, acid attack, cruelty against women, and kidnapping, etc.

Based on the NCRB data, the highest number of cases were registered in Uttar Pradesh, the country's most populated state (56,011). Maharashtra is a close second with 31,979 cases, followed by 30,992 in West Bengal, 29,778 in Madhya Pradesh, 25,993 in Rajasthan, and 23,082 in Assam. In Delhi, according to the NCRB report, 13,076 FIRs were registered in 2017, which is a decrease from 15,310 in 2016 and 17,222 in 2015. A ray of light comes from the states of Arunachal Pradesh, Goa, Himachal Pradesh, Manipur, Meghalaya, Mizoram, Nagaland, Sikkim, and Tripura where the registered crimes against women are only in three digits. This is not even 1% of all India. Similar trends can be noticed in the Union Territories, such as Chandigarh registered 453 cases, followed by 132 in Andaman and Nicobar Islands, 147 in Puducherry, 26 in Daman and Diu, 20 in Dadra and Nagar Haveli, and only 6 in the island territory of Lakshadweep. These figures from the data make one thing evident that one-half of the population continues to suffer disproportionately and faces the majority of the crimes committed. Women of all ages, religion, and caste continue to be the victims of sexual violence and heinous offenses ranging from rape, kidnapping, dowry-related deaths, physical and sexual assault to harassment at workplaces, abetment of suicide, and indecent representation of women in digital and print media.

These crimes act as barriers to their empowerment and self-respect in terms of gender. This puts a constant constraint on individual and societal welfare and has heavy social and economic costs. According to the World Bank report of 2018, violence against women is estimated to cost countries up to 3.7% of their GDP. It is an alarming figure as this amount is more than double of what most governments spend on education. Both the center as well as the state governments need to ensure that laws are enforced and perpetrators of physical, mental, or sexual violence do not remain unpunished. We also need more women representatives in the government. The presence of women stakeholders at all government levels leads to a better representation of their concerns in policy-making; it is also likely to lead to higher economic growth. The NCRB data is important because it highlights that the safety of women should continue to be an area of concern for the government. Various multipronged strategies, be it anti-discrimination and gender-based violence legislation, gender awareness campaigns, taking the recourse of judiciary, increasing system accountability, social perception, digital consciousness, strict action against child pornography, safe transport, hostels, safe roads and gender sensitization training, counseling, surveillance and increased crime-control policing, and such safe city programs may lead to lowering of the crime rates against women in society. On a positive note, some of the recent legislative, policy, and development measures laid by the current government have statistically improved laws and toughened transparency, and the effects can be seen in the little progress we are seeing in terms of reduction in the rate of crimes against women.

5.3 CRIME MAPPING AND GEOGRAPHICAL CONCERNS

Geographical concerns and events have both a spatial and temporal dimension. This means that they specifically occur in a particular context and timeframe. They might, for example, be a local community-based issue (such as a development proposal) that is a focus of people's attention for just a short period of time, or an environmental issue that affects the whole planet (for example, global climate change) which may be of concern for generations. A few contemporary geographical concerns have been listed below.

5.3.1 Population Distribution: High Population Densities

- Urban sprawl and pressures on the rural-urban fringe
- The sustainable city and diffuse city models
- Problems associated with high urban densities

5.3.2 The Interaction Between Transport and Location

- The currents of movement: Centralization and commuting
- Decentralization of industry
- Growth of second homes
- Problems associated with poor accessibility
- Issues associated with new transport projects

5.3.3 The Atmospheric and Climatic Framework: The Climate and Atmospheric Circulation

- The role of anticyclones
- Factors of the regional climate

5.3.4 Meteorological Risks: Cloudbursts and Flash Flooding

- Causes of floods
- Floods in the metropolitan drainage areas
- The effects of road works on river courses
- The effects of building on flood plains

5.3.5 Emissions and Air Quality: Dispersion of Air Pollutants

- Pollution sources
- The main air pollutants
- Air quality

5.3.6 Geology and Soils: Substrate and Relief: Geotechnical Problems and Risks

- Relief units
- Geotechnical issues
- Seismic activity
- Landslides and rock falls

5.3.7 Mining Activities: Mines and Restoration Work

- Observance of regulations
- Environmental impact of quarrying
- The restoration of exhausted quarries

5.3.8 Soils: Their Use and Pollution

- Soil types
- Soil vulnerability and degradation
- Assessment of soil pollution

5.3.9 Surface and Ground Water: The Water Supply Network: Characteristics and Socio-Environmental Problems

- Available resources and their use
- Water quality
- Flow volumes of surface waters

- Treatment processes
- Present and future water supply
- Water consumption and needs

5.3.10 Sewage: Types, Treatment, and Use

- The Catalan Water and Sewage Treatment Plan
- Generation and types of wastewater
- Wastewater treatment
- Use of treated waters
- Discharge of treated waters and sewer network

5.3.11 Landscape Ecological Structures and Continental Biology: The Environmental Landscape

- The creation of the landscape
- Types of landscape

5.3.12 The Biological Heritage: Fauna, Vegetation, and Urban Green Areas

- Vegetation
- Fauna
- Green areas

5.3.13 Nature Management and Environmental Parameters

- The environmental matrix
- Management of open spaces
- Forest fires
- Hunting and fishing
- Environmental impact studies (EIS)
- The problems of quantifying environmental phenomena
- Eco-landscape fragility index (EFI)

5.3.14 Energy: Energy Requirements: Strategy, Consumption, and Energy Infrastructures

- Energy management
- Energy consumption
- Energy generation
- Energy transformation
- Distribution networks
- The potential for the use of solid wastes and biomass
- Solar energy potential
- Wind and geothermal power

5.3.15 Geographical Concerns and Crime

Crime behavior varies with space and time. It is a dynamic phenomenon which needs a thorough introspection. A few geographical concerns and crime arousal have been discussed below.

5.3.16 Illegal Wild Animal Trafficking

The Interpol considers wild animal trafficking as the third-largest illegal business in the world after drug and arms trafficking. Such animal traffic raises a serious threat for the world's biodiversity survival. It is shocking to mention that the more endangered the species, the higher the price it is in the black market. The most endangered species are tropical birds (parrots, macaws, etc.), reptiles (serpents, crocodiles, etc.), arachnids (some types of tarantulas), monkeys (capuchins, chimpanzees, lemurs), and so forth. But animal trafficking is not just about selling those animals; there are also some serious concerns such as the sale of elephants or rhinoceroses ivory on the black market, used to make decoration items and/or in traditional Chinese medicine.

5.3.17 Illegal Lumbering

The uncontrolled logging to get precious wood for furniture or other goods or even for farm lands is one of the serious causes of this environmental crime. Illegal lumbering of sandalwood and medicinal herbs have been hyped in the past decade in the nation.

5.3.18 Wastage Dumping in Rivers and Lakes

This is one of the most dreadful crimes very common in the country most often caused by companies, factories, and common public. Chemical and toxic waste coming from factories is usually dumped in a controlled way, but this is not always the case. In these cases, waste is uncontrollably released into the environment, while at the same time polluting rivers, lakes, as well as land. This leads to not only affecting local wildlife but also, as a result of the water leaking into the soil, it finds its way to pollute the surrounding flora as well, affecting the food chain. There are many ways to avoid this waste-dumping problem, such as using sewage collectors or sewage plants, among others. Considering the effects that these crimes have on the environment, we have to bear in mind that many times they also involve people exploitation, corruption crimes and money laundering, killing (as in the case of illegal logging), and many others related.

5.3.19 Electronic Waste Mismanagement

The increased use of technology and IT has led to up to 50 million tonnes of electronic waste every year (computers, TV sets, mobile phones, appliances, etc.). Surprisingly, up to 75% of all of these wastes is estimated to leave the official circuit and a good deal of them to be illegally exported to Africa, China, or India. It is the case of Ghana's rubbish dump, a large electronic waste dump coming from the West.

5.4 NETWORK ANALYSIS

Network analysis (NA) is a series of interconnected strategies for representing actors' relationships and examining the social structures resulting from the recurrence of those relationships. The basic assumption is that the study of the interactions between individuals provides better explanations of social phenomena. This analysis is carried out by gathering organized relational data in matrix form. When actors are represented as nodes and their relationships as lines between node pairs, the idea of social network shifts from being a metaphor to an operational analytical tool that uses the mathematical vocabulary of graph theory and matrix and relational algebra (Fig. 5.7).

The deterministic approach emphasizes that NA allows analysis of how behaviors and attitudes are influenced by the social structure of relationships around an individual, community, or organization as structurally bounded purposeful acts may affect the social structure and vice versa. Network analysis can be regarded as a common methodological collection of techniques. Network theory is the study of graphs as a representation of either symmetric relations or asymmetric relations between discrete objects. In computer science and network science, network theory is a part of graph theory: a network can be defined as a graph in which nodes and/or edges have attributes. In many fields, network theory has applications including statistical physics, particle physics, computer science, electrical engineering, biology, economics, accounting, institutional analysis, climate science, ecology, public health, and sociology. Network theory applications include logistics networks, the World Wide Web, the Internet, gene regulatory networks, metabolic networks, social media, etc.

Under the term "combinatorial optimization," network problems involving finding an optimal way to do something are studied. Examples include network flow, shortest path problem, transport problem, transhipment problem, position problem, problem of matching, problem of selection, problem of packaging, problem of routing, critical path analysis, and PERT (Program Evaluation & Review).

A set of approaches to network analysis is also used in criminal intelligence to understand and act against serious crime occurrences, criminal groups, and criminal markets. These approaches are based in link analysis and increasingly include techniques from social network analysis. Network analysis techniques in criminal intelligence are used to organize data and reveal patterns in the nature and extent of relationships between data points. They also provide effective visualizations of both qualitative and quantitative data, which are valuable in presenting

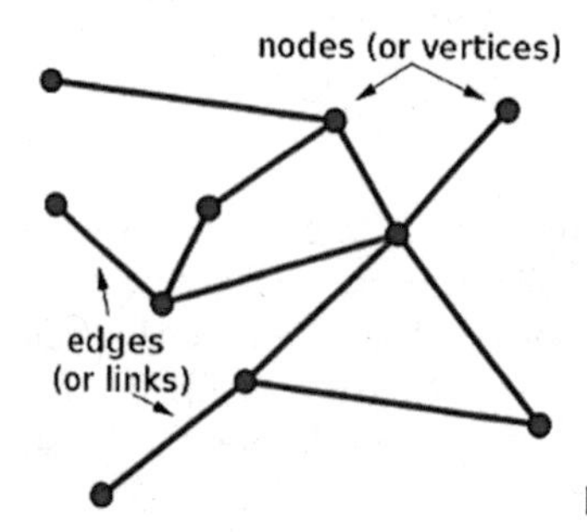

FIGURE 5.7 Nodes and Edges.

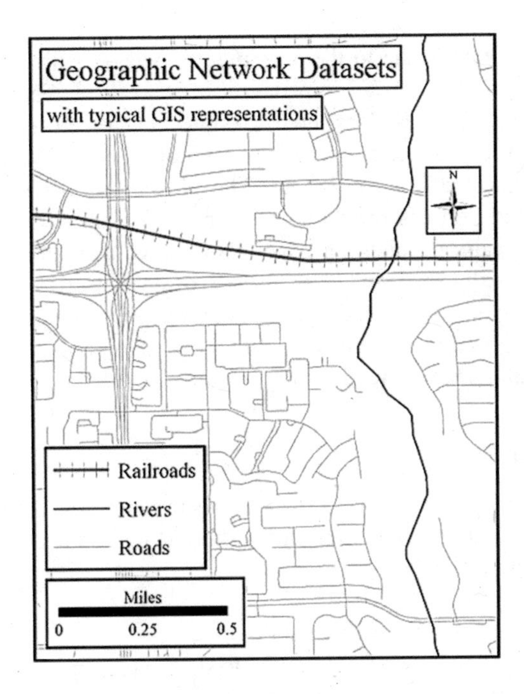

FIGURE 5.8 Geographic Network Database.

intelligence assessments. The link analysis methods used in criminal intelligence are a form of network analysis designed to discover and illustrate patterns in the connections between individuals, locations, organizations, objects, and events. Social network analysis has a tighter focus, concentrating on the relationships between people. Some social network analysis measures have utility in producing targeting recommendations for intelligence collection and operational disruption. The tools for network analysis, building networks, and creating and editing turns can be found in the Network Analyst Toolbox, which organizes these tools into three distinct toolsets. In GIS, the Network Analysis module is based on graph theory and topology which are sub-disciplines of discrete mathematical structures. A network is described as a set of vertices and edges, and graph theory addresses the description of networks, measures and matrices, and also comparisons of graphs and networks. Main topological properties that are addressed in graph theory are connectivity, adjacency, and incidence (Fig. 5.8). Analysis on networks is done on the basis of foresaid properties. Some common examples of network in GIS can be streets, power lines, and city centrelines.

5.4.1 Overview of Networks

Interconnected lines (known as edges) and intersections (known as junctions) form GIS networks to represent routes, upon which people, goods, etc., can travel. While traversing the network, an object follows the edges and whenever at least two edges intersect, junctions appear. Impedance can be defined as the increased cost of traveling in the network due to the attributes associated with junctions and edges. For an example of the foresaid impedance, a network representing roads can associate speed limits to the edges, and junctions can foreclose left turns. In graph theory, a graph is a structure amounting to a set of objects in which certain pairs of objects are "connected" in some way. The objects correspond to mathematical abstractions called vertices (also known as nodes or points), and each of the associated vertical pairs is called an edge (also known as path or line). A diagrammatic graph is usually represented as a series of dots or circles for the vertices. A graph can be directed or undirected, as do networks. In a directed network, only one way traveling is permitted whereas in undirected there is no such compulsion (i.e., travel can happen in both directions).

5.4.2 Shortest Path

The most common case of Network Analysis is to determine the shortest path from a source to a destination. The criterion for shortest path in a network of streets can be of different kinds, such as

- Distance from source to destination, or
- Time required to reach from source to destination, or
- Monetary value (i.e., cost associated from reaching source point to destination point).

For example, a person driving a vehicle may look for the shortest path based on time value (i.e., the path that can take him/her in minimum amount of time from his/her current location to the destined required facility). The shortest path problem in graph theory is the problem of finding a path in a graph between two vertices (or nodes) so that the sum of the weights of its constituent edges is minimized. The problem of finding the shortest path between two intersections on a road map can be modeled as a special case of the shortest path problem in graphs, where the vertices correspond to intersections, and the edges correspond to segments of the lane, each weighted by segment length. The shortest path (A, C, E, D, F) between vertices A and F in the weighted directed graph (Fig. 5.9).

The problem is also sometimes called the shortest path problem for a single pair, to differentiate it from the variants that follow:

The shortest single-source path problem, where we need to find the shortest paths from the source vertex v to all other vertices in the graph.
The shortest single-destination path problem, in which we have to find shortest paths from all vertices in the directed graph to a single destination vertex v. This can be reduced to the shortest single-source path problem by reversing the arcs in the directed graph.
The shortest path problem for all pairs, in which we have to find shortest paths between each pair of vertices v, v in the graph.

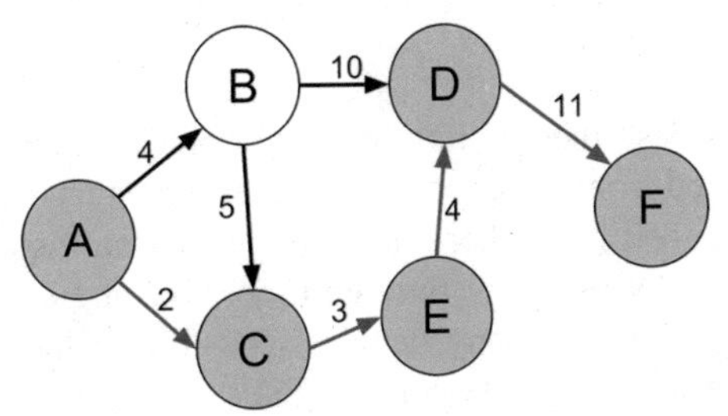

FIGURE 5.9 Example of Shortest Path.

These generalizations have significantly more efficient algorithms on all relevant pairs of vertices than the simplistic approach of running a single-pair shortest path algorithm.

Shortest path algorithms are used to find directions automatically between physical locations, such as driving directions on online mapping websites such as MapQuest or Google Maps. There are fast specialist algorithms available for this application. If one views a non-deterministic abstract machine as a graph in which vertices describe states and edges describe possible transitions, shortest path algorithms can be used to find an optimal sequence of choices to reach a particular target state, or to set lower limits on the time needed to reach a given state. For example, if vertices represent the states of a puzzle like a Rubik's Cube, and each directed edge corresponds to a single move or turn, the shortest path algorithms can be used to find a solution using the minimum number of moves possible. This shortest path problem is sometimes referred to as the min-delay path problem in a networking or telecommunications perspective and typically related to a broader path problem. For example, the algorithm can check for the shortest (min-delay) broadest path, or the widest (min-delay) shortest path. Certain uses, frequently studied in operations research, include the architecture of plants and facilities, robotics, transport, etc.

5.4.2.1 Road Networks

A road network can be considered as a graph with positive weights. The nodes represent road junctions and each edge of the graph is associated with a road segment between two junctions. The weight of an edge may correspond to the length of the associated road segment, the time needed to traverse the segment, or the cost of traversing the segment. Using directed edges it is also possible to model one-way streets. Such graphs are special in the sense that some edges are more important than others for long-distance travel (e.g., highways). This property has been formalized using the notion of highway dimension. There are a large number of algorithms that exploit this property and are therefore able to calculate the shortest path much faster than in general graphs would be possible. Such algorithms all work in two phases. The graph is preprocessed in the first iteration, without knowing the source or target node. The second phase is the phase of your query. Source and destination node are known at this point. The concept is for the road network to be static, so the pre-processing method can be performed once and used on the same road network for a large number of queries.

5.4.3 Traveling Salespeople

The problem of traveling salespeople is described as hitting every point in a network as efficiently as possible. It is derived from the concept of a salesperson trying to reach a targeted collection of cities to sell his or her goods in the easiest, most efficient way possible, either through money made, or through time. UPS uses an algorithm for traveling salespeople to deliver as many items as possible to their customers every day. The problem of a traveling salesperson (TSP) asks the following question: "Given a list of cities and the distances between each pair of cities, what is the shortest route possible that visits each city and returns to the city of origin?"

The problem was first formulated in 1930 and is one of the most intensively studied optimization problems. It serves as a benchmark for many methods of optimization. Even though the problem is computationally difficult, many heuristics and exact algorithms are known, so that some instances with tens of thousands of cities can be solved completely and even problems with millions of cities can be approximated within a small fraction of 1%. Even in its purest formulation, the TSP has several uses, such as preparation, storage, and microchip manufacturing. Slightly modified, it occurs in many places as a sub-problem, including DNA sequencing. For example, in these applications, the concept city represents consumers, soldering points, or DNA fragments, and the concept distance represents travel times or cost, or a measure of similarity between DNA fragments. The TSP often appears in astronomy, because astronomers will want to reduce the time spent shifting the telescope between the sources when observing other sources. Extra limitations such as limited resources or time limits can be implemented in many applications.

The roots of the problem of the traveling salesperson are unclear. A 1832 handbook for traveling salespeople discusses the problem and provides example tours across Germany and Switzerland, but it does not include any mathematical treatment. The problem of the traveling salesperson was developed mathematically by the Irish mathematician W.R. in the 1800s. Icosian Game by Hamilton was a leisure puzzle based on finding a Hamiltonian loop. TSP can be modeled as an undirected weighted graph, so cities are the vertices of the graph, paths are the edges of the graph, and width of a path is the weight of the edge. Starting and finishing at a given vertex after visiting each other vertex exactly once is a minimization problem.

The model is often a complete graph (that is, each pair of vertices is connected by an edge). If there is no route between two cities, adding an arbitrarily long edge completes the diagram without affecting the optimal trip. In the symmetric TSP, the distance in each opposite direction between two towns is the same, forming an undirected graph. The symmetry reduces the number of possible solutions by half. In the asymmetric TSP, paths in both directions may not exist, or distances may vary, creating a directed graph. Traffic collisions, one-way streets, and airfares are examples of how this pattern could break down for towns with different departure and arrival fees.

Symmetric TSP with four cities-

The Vehicle Routing Problem (VRP) is a problem of combinatorial optimization and integer programming that asks "What is the best set of routes for a vehicle fleet to traverse to deliver to a given set of customers?". This generalizes the well-known

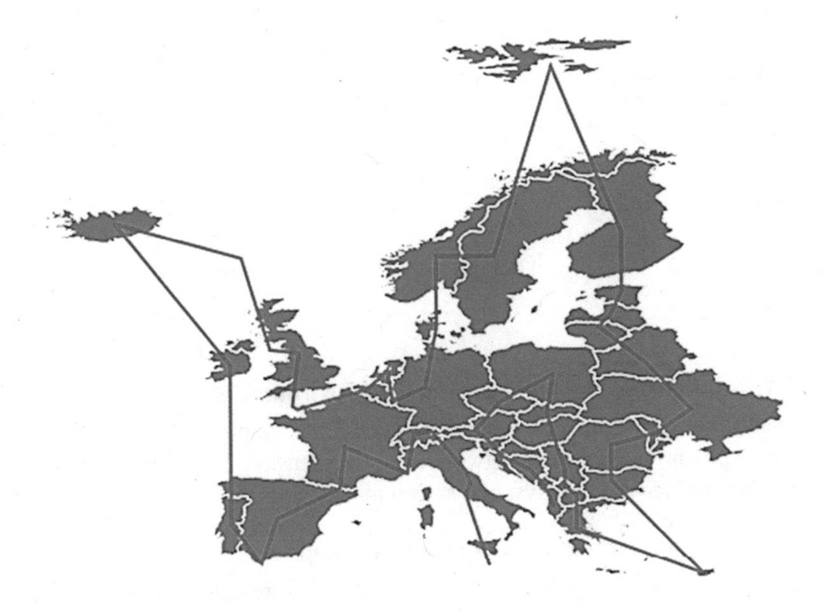

FIGURE 5.10 The Traveling Salespeople Network.

problem of a voyaging salesperson (TSP) (Fig. 5.10). It first appeared in a 1959 paper by George Dantzig and John Ramser, in which the first algorithmic solution was written and applied to deliveries of petrol. The background is often that of delivering goods to consumers who have placed orders for such products located at a central depot. The VRP's goal is to reduce overall cost of the road.

5.4.4 Network Partition

Network partitioning is a division of regions into zones or subcategories in a network. Such regions are sized based on a network's proximity to specific points (Fig. 5.12). This is popular in metropolitan-area fire stations. A network partition refers to the decomposition of the network into relatively independent subnets for separate optimization as well as the separation of the network due to network system failure. In both cases, subnet partition tolerant behavior is expected, which means it still works correctly even after the network is partitioned into multiple subsystems. For example, if the network switch system between the two subnets fails, a partition will occur in a network with several subnets where nodes A and B are located in one subnet, and nodes C and D are located in another. Nodes A and B can no longer communicate with nodes C and D in that case, but all nodes A–D work the same way as before (Fig. 5.11).

5.4.5 Network Analysis Workflow

There are basic steps you need to take in order for you to conduct a network analysis in a GIS system. Each segment goes through the general process you need to do before addressing network analysis issues.

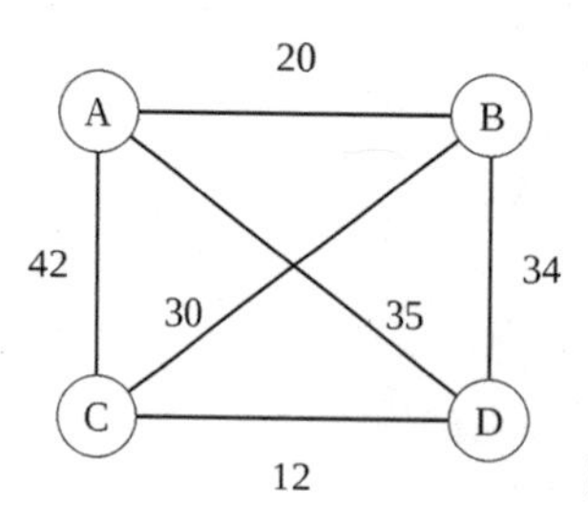

FIGURE 5.11 Symmetric TSP with Four Cities.

Step 1: Configuring the Network Analyst Environment

An analyst network is an extension to ArcGIS. Thus, before performing any network analysis, you must activate the Network Analyst extension. You will also need to view the toolbar for Network Analyst and screen the Network Analyst window from there.

Step 2: Adding a Network Data Set to ArcMap

You need to have a network on which to conduct the research before you can carry out a network analysis. So your next step is to add a dataset layer for the network to ArcMap. If the network is not built, then you will need to build it. If the source features have been modified, or the network attributes that apply to the source features have changed, the network data set needs to be rebuilt.

Step 3: Creating the Network Analysis Layer

The inputs, properties, and results of a network analysis are stored in a network analysis layer. It contains a workspace in memory with classes for network analysis for each input type, as well as for the results. The functions and documents inside the groups of network analysis are referred to as artifacts of network analysis. Some network analytics layer properties allow you to further define the problem you wish to solve. On network databases, network analyses are always performed. A network analytics layer must therefore be connected to a network data set. If you use a geoprocessing method to build a network analysis layer, you set the network data set to be a function parameter. In ArcMap, a network data set must be added first so that Network Analyst can connect the analytics layer to the network data set when constructing an analysis layer.

Step 4: Adding Network Analysis Objects

The objects of network analysis are features and documents that are used as input and output during network analysis. Definitions include exits, roads, barriers, and services. You can connect objects to input classes for network analysis, but you won't be able to add them to output-only classes. Objects for the output-only network analysis can be generated only by the solver. For example, the Route class is only an output in a route analysis layer, so that only the solver can create route objects. There are different ways to add objects into classes. The two most frequently chosen choices are: one, loading multiple functions at once into a network analysis class, and two, inserting one object at a time interactively.

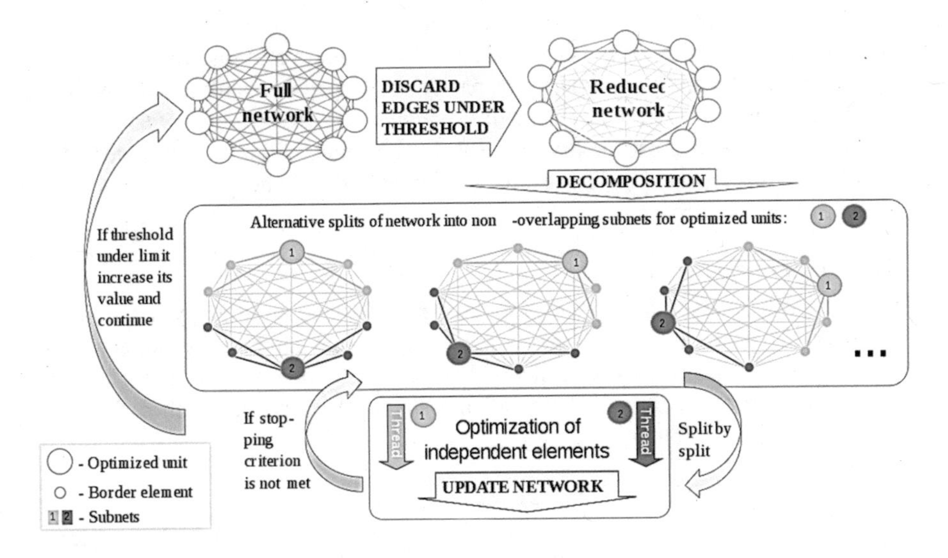

FIGURE 5.12 Network Partitioning.

Step 5: Setting Network Analysis Layer Properties

The network analytics framework also has properties which are more general to the application than those of the subjects of its network analysis. The general properties of the application are the attribute of network impedance to use, the attributes of the constraint to obey, and so on. Additionally, the properties are unique to the type of analysis being carried out. The analysis layer's Layer Properties dialog box provides access to those properties.

Step 6: Performing the Analysis and Displaying the Results

Once you have built your layer of analysis, inserted objects for the input network analysis, and set the parameters for the objects of analysis and the layer of analysis, it's time to resolve the network problem.

5.4.6 Types of Network Analysis Layers

A Network Analyst can find the best way of going from one place to another or visiting multiple locations. Positioning points on the screen, entering an address, or using points in an existing feature class or feature layer will define the locations interactively. If you have more than two places to visit, you will determine the best route for the user-specified order of locations. Additionally, Network Analyst can determine the best sequence to visit the sites, known as solving the problem of traveling salespeople. Knowing the nearest hospital to an accident, the nearest police vehicles to a crime scene, and the nearest shop to a customer's address are all examples of nearest facilities concerns. You can specify how many to find when you find the nearest facilities, and whether the direction of travel is to or away from them. Once you have found the nearest facilities, you will be able to display the best route to or from them, return travel costs for each route and display directions to each facility. In addition, you can specify an impedance cutoff that Network Analyst should not search for a facility beyond. For example, you can set up a problem with the nearest facility to check for hospitals within a 15 minutes' drive time from an accident site. The results will not include any hospitals that take more than 15 minutes to reach.

5.4.6.1 Service Areas

For Network Analyst, the coverage areas can be located around any location on a network. A service area for the network is a region that encompasses all accessible streets; that is, streets within a defined impedance. For example, the 10-minute service area for a facility includes all the streets that can be reached from that facility within 10 minutes.

5.4.6.2 OD Cost Matrix

Using Network Analyst, you can build a cost matrix from multiple sources to multiple destinations for the origin-destination (OD). An OD cost matrix is a table containing impedance of the network to each destination from each origin. Additionally, it ranks the destinations to which each origin connects in ascending order based on the minimum network impedance required to travel to each destination from that origin. For each origin-destination pair, the best network route is found, and the costs are stored in the output line attribute table. Even if the lines are straight for performance reasons, the network cost is always stored, not straight-line width. The graph shows the results of an analysis of the OD cost matrix, which was set to find the cost of reaching the four nearest destinations from each origin.

5.5 TRANSPORTATION MODELING

For Transportation Planning, basic functions already existing for GIS including buffer, overlay, question, etc., are useful. Nonetheless, for planning purposes, deeper analysis of the network data is available. Examples of such higher uses include network flow equilibrium models, travel requirement models, and trip generation and delivery, as well as activity-based models and interaction models for transportation/ land use. The latter use is especially useful as demand for transportation affects land use and, reciprocally, the effect of the changed land use on transportation. There is limited commercial software for performing these tasks, but professional programmers should be able to design programs for performing those tasks where software is unavailable or non-existent. The word "models" is widely used in transport planning. This concept is used to refer to a collection of mathematical equations used to describe how people make choices while traveling. Travel demand comes about as a result of thousands of individual travelers making decisions about how, where, and when to travel (Fig. 5.13).

These decisions are influenced by many factors, such as family situations, the characteristics of the person making the trip, and the choices available for the trip (destination, route, and mode). Mathematical structures are used in making certain choices to reflect (model) human behavior. Models require a set of assumptions to

FIGURE 5.13 Transportation Model.
Source: Center for Urban Transportation Studies University of Wisconsin.

work, and the data available to make predictions are small. The model's coefficients and parameters are set (calibrated) to match existing data. Such partnerships are usually believed to be true and will continue to be constant in the future. Modeling of travel demand was first developed in the late 1950s as a means of planning the highways. As the need emerged to look at other problems such as transport, land use concerns, and air quality analysis, the modeling process was updated to add additional techniques to address these issues.

Models are important because transport plans and investments are dependent on what the models say about traveling in the future. Models are used to estimate the number of journeys that will be made on an alternative transport system at some future date. These projections form the basis for transport plans and are used in major investment analyses, environmental impact statements, and investment goals. Models are based on theory of how travel happens. It is important to have a clear understanding of the modeling process to help clarify transport plans and their recommendations. The method of travel forecasting is at the very heart of urban transport planning. Travel forecasting models are used to predict potential traffic and are the basis for determining the need for new road infrastructure, improvements in transit service, and changes in policies and trends on land use. Modeling travel demand involves a series of mathematical models trying to simulate human behavior while traveling.

The simulations are performed in a sequence of steps referring to a set of traveler decision questions. In response to a given system of highways, transportation, and policies, attempts are made to simulate all the choices that travelers make. There are many observations that need to be made about how people make choices, the variables that they weigh, and how they respond to a specific option for transport. The travel simulation method follows trips as they start in a zone of trip generation, pass through a network of links and nodes, and end in a zone of attraction. The simulation process for the four basic models used is known as the four-stage process. These are generation of trips, distribution of trips, split mode, and assignments to traffic. Such templates are used to answer a number of questions, as discussed in the rest of the priming. In addition, it will also describe the process used to represent urban areas and the use of model results.

Trip generation is the first step in travel forecasting. In this phase, land use statistics, population, and economic projections are used to predict how many individual trips to and from each zone will be made. This is done separately by purpose of trip. Trip uses which can be used include home-based work trips (work trips beginning or finishing at home), home-based shopping trips, home-based other trips, school trips, non-home-based trips (trips that neither start nor end at home), truck trips, and taxi trips. Trip generation uses trip rates that are an average for a large study area section. Trip productions are based on household features, such as the number of people in the household and the number of available vehicles. For example, it might be presumed that a household of four people and two vehicles produces 3.00 work trips a day. Trips per household will then be generalized to journeys per state. Trip attractions usually depend on the level of work in a region. For example, for every person working in that zone, a zone could be expected to attract 1.32 home-based work trips. The generation of trips is used to measure the trips for the user. These are later modified for deciding vehicle journeys in the split/auto occupancy stage process.

5.5.1 Common Limitations and Issues

The following are some of the common limitations that might be of concern in the generation of trips.

Decisions are independent: Travel behavior is a complex process in which one household member's decisions are often contingent on others within the household. Child-care needs, for example, may influence how and when people travel to work. This interdependence for the making of trips is not taken into account.

Trip purposes are limited: A simpler trip pattern results in no more than four to eight trip finishes. All shopping trips are treated the same whether grocery or lumber shopping is done. Home-based "other" trip purposes cover a wide range of purposes (e.g., medical, visiting friends, banking, etc.) that are affected by a broader range of factors than those used in the modeling process.

Limited variables: Trip making is seen as a function of just a few variables including auto ownership, household size, and employment. Generally, other considerations such as the nature of the transit service, ease of walking or cycling, fuel prices, land use design, and so on are not included.

Combinations of trips (trip chaining) are ignored: Travelers can often combine a variety of purposes into a sequence of trips as the errands run and connect activities together. This is called the chaining of journeys, and is a complex process. The modeling process treats such combinations of trips in a very limited manner. Non-home-based journeys, for example, are calculated based only on work characteristics of zones and do not consider how members of a household manage their errands.

Feedback, cause-and-effect problems: Often trip generation models measure trips as a function of variables that might in effect depend on how many trips there are. For instance, shopping trip attractions are seen as a feature of retail jobs, but it could also be argued that the number of retail employees in a shopping center may depend on how many people come to shop there. The question of "chicken and egg" often arises in travel predictions, and is difficult to avoid. Another example is that traveling depends on auto availability, but it could also be argued that the number of cars owned by a household will depend on how involved they are when making travels. Trip generation only finds the number of trips that begin or end at a particular zone. Through the cycle of trip distribution these trip ends are joined together to create an origin-destination pattern for trips. Trip distribution is used to represent the process of destination choice (i.e., "I need to go shopping but where should I go to meet my shopping needs?"). Distribution of trips leads to a substantial increase in the amount of data that needs to be addressed.

Quite wide origin-destination tables: For example in its O-D table a 1,200-zone study area would have a potential 1,440,000 trip combinations. For each intent of the trip, separate tables are also done. The gravity model is the most widely used trip, distribution method. The gravity model takes the journeys produced in one zone and distributes them to other zones based on the size of the other zones (as measured by their travel attractions) and the distance to other areas. A zone with a large number of travel attractions would earn more dispersed travels than one with a small number of travel attractions. The other element used in the gravity model is distance from the possible destinations. The number of trips to a given destination decreases with the distance (it is inversely proportional) to the destination. The distance effect is discovered through a calibration method that seeks to lead to a similar distribution of trips from the model, as observed from field data.

Transportation models are called upon to provide predictions for a diverse set of issues that may go beyond their capabilities and original purpose in some cases. Travel demand management, trip reduction programs focused on the workplace, pedestrian and cycling systems, and land use policies may not be treated well in the process. Models of transport travel forecasting use packaged computer programs that have limitations on how easily they can be changed. In some cases, the models can be updated to fit additional factors or procedures (quick fix) while significant changes are required in other cases or a new software is needed. The following are some possible model modifications that may help to improve their usefulness:

All models are based on travel patterns and conduct data: If that data is obsolete, the findings will be incomplete or incorrect, no matter how good the models are. One of the most important ways to improve the quality and reliability of the model is to have a good basis for using recent data to calibrate the models and to ensure that their accuracy is tested. Models need to prove that they provide an accurate picture of current travel before using them to predict future travel. An improved representation of bicycle and pedestrian travel is needed. In modern travel, demand models commuting by bicycle and walking is not done well. Better approaches are needed to deal with these types of journeys. This can be done by incorporating factors in models of trip generation that relate the making of trips to the amenities of the pedestrian or bicycle. These types of trips could also be expanded to include mode choice methods.

Better auto occupancy models: Current practices for auto occupancy appear to be insensitive to a wide range of policies which may contribute to more or less carpooling. Auto-occupancy policies need to be responsive to the cost of parking and travel costs as well as the number of trips that occur between a destination and origin. Also, it may be desirable to treat ride sharing among family members differently than carpooling between persons from different households. Procedures that increase the number of trip purposes to deal with segments of the market that are likely to share trips may help with this issue.

Better time of day factors: Congestion rates in hours other than the peak period are required to gain a better understanding of the essence of congestion as it persists throughout the day and into the future over time. Hourly conversion factors must be scrutinized very carefully to ensure that they represent actual traffic variations.

Use more trip purposes: Additional travel purposes (market segments) may provide a way to better represent complex household travel patterns and chaining trips. That would also provide procedures for the generation of trips that are sensitive to more factors that would result from travel management techniques.

Better representation of access: The transport models are not well reflected in land use policies that promote transit use or provide high-quality site design with good pedestrian access. Improved methods are needed to measure the disutility of the transit and highway travel access portions. These approaches would include calculating a sensitive access index in areas that used more bus/pedestrian/bike friendly architecture for easy access and waiting for transit vehicles.

Incorporate costs into trip distribution: Trip distribution models should use a generic distance calculation that involves travel costs by various means including cost of parking. Therefore, these models will help show the sensitivity of travel patterns to cost changes.

Add land use feedback: It is important to take measures to close the forecasting process loop in order to enable a better representation of the land use and travel demand interactions. Simulation of land use models should be applied to the model series to help determine how a new transport system can lead to changes in land use.

Add intersection delays: Most congestion is experienced in an urban traffic network at traffic signals or stop signs rather than on intersection roads. Travel forecasting models should include routines that calculate the intersection delay encountered. Furthermore, splits of the intersection signal should be viewed as a variable that would be changed as the cycle of traffic assignment iterates to reach equilibrium.

5.5.2 Vehicle Routing Problem

A dispatcher managing a vehicle fleet is often required to make the vehicle routing decisions. Another such decision is how best to assign a group of customers to a vehicle fleet, and how to arrange and plan their visits. The aim of solving these Vehicle Routing Problems (VRP) is to provide a high level of customer service by accommodating any time frames while keeping the total operational and investment costs as low as possible for each route. The limitations are to complete the routes with the resources available and within the time limits placed by driver work shifts, driving speeds, and customer commitments. Network analysis offers a problem-solver for vehicle routing, which can be used to find approaches for such complex fleet management tasks. Consider an example of delivering goods from a central warehouse location to the grocery stores. There is a three-truck fleet at the factory. The warehouse works only within a given window of time from 8:00 a.m. to around 5:00 p.m. when all trucks have to return to their warehouse. That truck has a 15,000- pound capacity, which limits the amount of goods it can carry. Every store has a requirement for a certain quantity of goods (in pounds) that must be shipped, and each store has time periods that restrict when deliveries are due. In addition, the driver can work just 8 hours a day, needs a lunch break, and is charged for the amount spent on driving and servicing the shops. The purpose is to create an itinerary for each driver (or path) so that deliveries can be made while complying with all the service requirements and reducing the driver's total time spent on a particular route. The following figure shows three routes obtained by solving the aforementioned vehicle routing problem.

5.5.2.1 Location-Allocation

Location-allocation helps you select which facilities to operate from a set of facilities based on their potential interaction with the demand points. It can help you answer such questions as:

In view of a set of existing fire stations which site would provide the best response times for the community for a new fire station?
If a retail company has to scale down, which stores should it close to retain the most competition overall?
Where will a plant be designed to reduce the gap to distribution centers?

Facilities would represent the fire stations, retail stores, and factories in these examples; demand points would represent buildings, customers, and distribution centers.

5.5.2.2 Time-Dependent Analysis

All of the solvers allow you to integrate live and historical traffic data into an analysis so that you can find the best route for a given time of day; determine the best place to plan an ambulance at 8:00 a.m., 12:00 p.m., 4:00 p.m., etc.; and create service areas for different times of day. With various dates and times, the outcomes of any study may vary, because traffic conditions and travel times can change.

Concluding Remarks

Micro-level investigations and real-time analysis proves that demography affects crime, catering to the various parameters of intervention such as impact of age, sex, occupational structure, and society on criminal behavior, and the difference between compositional and contextual effects of demographic structure on aggregate crime rates. The intersection of criminal and demographic events in the life course; the influence of criminal victimization and aggregate crime rates on residential mobility, migration, and population redistribution are some pertinent issues of analysis. Geographical concerns and events have both a spatial and temporal dimension highlighting the approaches of network analysis used in criminal intelligence to understand and act against serious crime occurrences, criminal groups, and criminal markets. Transportation models, travel demand management, trip reduction programs, and models of transport travel forecasting are also pertinent features of geospatial analysis in crime investigations.

REFERENCES

Conrad, T. and Irene, B. T. (1958) *The Changing Population of the United States*. John Wiley and Sons, New York, London, 12–16.

Donald, J. B. (1969) *Principles of Demography*. Wiley, 32.

Gordon, A. C. (2016) *Fundamentals of demographic analysis: Concepts*, measures and methods. Springer Series on Demographic Methods and Analysis, 38(2).

Hervé, L. B. (2008) *The Nature of Demography*. Princeton University Press, Social Science, 361.

Pathak, L. (1998) *Population Studies: The Discipline, Development Pattern and Information System*. Rawat Publications, 17–18.

Scott, J. S. and Steven, F. M. (2000) Crime and Demography: Multiple Linkages, Reciprocal Relations. *Ann. Rev. Sociol.* 26, 83–106.

Scott, J. S., Katherine, T., and Sunita, B. (2014) Skewed sex ratios and criminal victimization in India. *Jastor*, 51(3), 1019–1040.

Guttentag and Secord. (1983) *Too Many Women*. Sage Publications, 277.

6 Mapping for Operational Police Activities

An understanding of change in recent years is essential while planning for the future. The ways in which our present society is changing affects the future of policing in dynamic ways. These changes are being shaped by social, economic, and political factors that are substantially beyond the immediate control of politicians and policy-makers. There has been a tremendous increase in the incidence of crime since the 1950s in all sections of the society. Further, the prevailing explanations of that increased incidence, the availability of relatively anonymous and easily disposable property together with declining informal social controls suggests that the long-term impacts will not be easy to reverse. By contrast with property crime, however, the rise in violent crime over a given period has been much less acute.

There has been a growth in social concerns about crime and fear of crime such that law and order has become a major public policy issue and, therefore, a political issue. Parallel with it there has been an increased awareness of the risk of becoming a victim of crime as well. In addition, private security services are also developing rapidly and recent years have seen the emergence of voluntary, self-help organizations providing protection for individual as well as local communities. Though, this has not satisfied public demand and it is not at all clear how the desire for increased security or the reduction of risk, could fully satisfy the society in future.

6.1 POLICE PATROLLING AND SURVEILLANCE

Patrol is well known as the backbone of every police department. The basic philosophy and strategy of preventive patrol has not much changed from Peel's time: The patrol officer makes circuits in a specified area, often called a beat. During Peel's time, most patrols were done on foot, with the occasional horse patrol. Technology ushered in the automobile, and modern police forces now take full advantage of the benefits offered by movement vehicles. The most important of these advantages is the specified area that a single officer can cover. The effectiveness of patrolling operations within a department are usually judged by three major functions: They include answering calls for service, deterring crime by a highly visible police presence, and investigating suspicious circumstances, out of which crime deterrence is the most controversial.

6.1.1 The Kansas City Preventive Patrol Experiment

In the 1970s, criminal justice researchers began to question the underlying assumption of preventive patrol process. They designed an experiment to find out if preventive patrol reduced crime and possibly made citizens feel safe from crime. In other words, they wanted to investigate if the number of officers on patrol in a given area have an impact on both actual crime and citizens' perceptions of crime (The Kansas City Preventive Patrol Experiment, 1974).

The researchers' experiment was conducted in concurrence with the Kansas City, Missouri, Police Department. The department divided the city into 15 beat areas and into three groups. To provide the experimenters a needed controlled group environment, one cluster of five beats made no changes in the amount of patrol officers working in the area. Whereas in the second area, the police withdrew all preventive patrol and served a completely reactive role. They entered this "volatile" area only when calls for service were received. In the third area, they raised preventative patrol to four times their normal level. If the conventional wisdom about the effectiveness of preventive patrol was held true, then the experimenters observed a higher crime rate in the volatile reactive area, no change in the crime rate in the control area, and a drop in the intensified patrol area.

The researchers were found astonished to observe almost no difference in actual crime or citizens fear of crime. Citizens' opinions about how good a job the police were doing did not really change. It seemed that law-abiding citizens and criminals alike simply did not notice the changes in their efforts. As one would expect, this caused a flurry of opinions to rise regarding the interpretation of these findings. Some argued that the findings were wrong, and that preventive patrol was and always had been a good positive thing. Others argued that patrol was just a bad idea and that the police should focus on other different things. Many stood the middle ground, focusing on making patrolling more effective by changing the way it was done. One of the few things that almost all commentators agreed on was that just adding up more officers out on the street would have little impact on crime. What was needed was a fundamental change in the patrolling and policing profile.

6.1.2 The Proactive Paradigm Shift

While the research evidence seems to indicate that the mere presence or addition of uniformed officers in an area does little to deter crime, the same cannot be said for more aggressive patrol strategies (Hoover, 2014). Proactive patrol operations have shifted from random to targeted cases. Specific types of offenders, specific places, and specific types of victims are now being considered. Myriad tactics fall under this general philosophy which discusses undercovering of operations, the use of informants, using decoys, saturating problem areas, and frequent patrols of hot spots are just a few examples. An important argument about how to better utilize patrolling services is that random patrols do not work well because crime is not a random phenomenon. While it may seem fair enough to give every neighborhood in a city an equal amount of police time and resources, it is astonishingly inefficient. A smarter and justifiable use is to concentrate police resources in high-crime areas, and limit resources in areas that experience very little crime. Research evidences suggests that this strategy indeed has a positive impact on crime control. Other strategies, such as those used in the San Diego Field Interrogation Study, have indicated that aggressively interrogating suspicious persons can lead to a reduction in both violent crime and disorder. The New York City Street Crimes Unit has had success using decoys to apprehend repeat offenders. By having an undercover officer play a "perfect victim," officers were able to increase dramatically arrests of attackers.

6.1.3 Problem-Oriented Policing

The traditional model of policing in the developed nations has always been reactive in nature. The primary methods used by policing departments were preventive patrols and retroactive investigations. Initial efforts at innovation were designed to be proactive, but they focused on the deterrence of crime through a limited "toolbox" of arrests, summons, and citations. But there has been a decadal evolution in this context, with a shift in its focus, with the confluence of two major developments in how both practitioners and academics view policing. The first being Problem-Oriented Policing (POP) and the other with a broader philosophy including POP, known as Community-Oriented Policing (COP).

The concept of problem-oriented policing initiated with an article published by Herman Goldstein in 1979, where he suggested that the basic and most fundamental job of the police was to deal with community problems. To undertake this task effectively, the police needed to develop a much larger toolbox, and a much more sophisticated method of detecting, analyzing, and ultimately solving such pertinent issues. A major tool in the analysis of community problems is the *Problem Analysis Triangle* (Palmiotto, 2000). The idea of a crime triangle is to graphically depict the interaction between the features of the victim, the features of the location, and the features of the offender geographically. As Spelman and Eck in 1989 pointed out, 10% of crime victims are involved in up to 40% of victimizations, 10% of offenders are involved in 50% of crimes, and around 10% of addresses are the location for about 60% of crimes (Philip et al., 2000). This concept suggests that a focus on a few high-volume victims, offenders, and locations can maximize the impact of shortage of police resources. To understand the problem-solving process, it is essential to understand a problem and its scope, which interests the police mission.

6.2 CRIME SCHEDULING AND TIME MANAGEMENT

There have been several attempts to deter, detect, and respond to crime. There has been an explicit paradigm shift in the ways we have been addressing criminal issues. Cities in the developed nations have started to embrace the concept of real-time crime monitoring centers. These centers are designed to gather, interpret, and disseminate information to enhance situational awareness of events as and when they occur with respect to exact time. As advances in technology continue to provide improved but more complex sources of data, having a centralized location to interpret and manage this information becomes critical.

Various software programs have been designed to integrate multiple data sources into meaningful intelligence. Programs typically receive data from video surveillance, license plate readers, vehicular GPS tracking, social media data streams, geographical information systems, and local crime statistics acquisition centers. Some also incorporate systems for gunshot detection, radiation and chemical sensory, facial recognition, and predictive policing have come up recording incidences temporally. Because visualization is one of the most beneficial methods for achieving situational awareness, video monitoring is a critical component of these systems. A city will initially identify its areas of greatest concern, and then install multiple video surveillance monitors in these locations. Such areas include those with high crime rates, high profile and critical infrastructure, or frequent mass gatherings based on past data. If these areas are later flagged for abnormal activity or if first response units are dispatched for calls involving crime or violence, all surrounding cameras are immediately activated to enhance monitoring capabilities, thus recording and maintaining time schedules of past incidences help in the future monitoring of crime. The videos are used to obtain real-time visual information of an active incident and/or be later reviewed to obtain additional or detailed video evidence. Some systems are even capable of allowing users to flag and pinpoint specific identifiers so that on review, the video will skip forward to only show screens that include the identifier (e.g., red car). While city cameras are a key source of the feed, many of the programs also integrate public monitoring systems that are owned by private citizens and businesses in a region. Some cities have a requirement for certain businesses to participate in the monitoring program, while others have an optional enrollment.

While the focus of real-time monitoring systems tends to be on crime control, prevention, and suspect apprehension, there is great potential to use this software to improve mass gathering monitoring and mass casualty response. Appropriately configured video monitoring systems can be used in a city to enhance the crowd surveillance capabilities of public safety officials working on the ground at public events. And when a mass casualty incident does occur, the system can be used to more accurately and rapidly determine the impact of the incident and resources/personnel needed to respond. More complete information can be relayed to dispatch and first responders prior to their arrival on scene and early notifications and timely patrolling. These centers and systems are not without limitations or criticism, but arguably have great potential to enhance our prevention and response capabilities for crime, mass casualty, and terror events.

6.3 STRATEGIC CRIME ANALYSIS IN AJMER CITY (RAJASTHAN), INDIA: ANALYZING THE UNDERLYING DRIVERS OF CRIME USING GEOSPATIAL TECHNIQUES

Crime is an exercise done by an individual or a group of persons that is unethical, harmful, and unsocial to the society. Crime can be the consequence of illiteracy, poverty, or revenge. Such crime mapping is done physically to help trace suspects and establish their modus operandi in most of the states in the country. It helps the police determine areas where a particular group of suspects or individuals are active. For such analysis, it becomes mandatory to have a thorough introspection about the concerned geographical region. A wide area of socio-economic variables are classified including household income, educational achievement, employment status, and poverty status; other demographic variables, age, religion, and race, are analyzed. The relation between these variables and crime type provides understanding of crime in both temporal and spatial contexts.

GIS has been gradually picking up as an important technology to the social crime prevention systems in India. The present GIS provides an opportunity for crime analysis and generates inputs towards crime prevention and planning. The spatial research facilitates better community policing rather than analytical prediction. Crime mapping using GIS tools provides basic criminal investigative analysis that includes activities such as geographic profile and specific case-support-based crime investigation. Such GIS-enabled crime maps provide law and order force to assist in strategic planning, crime analysis, and operations.

It is very important to understand that geography of an area plays a dominating role in crime sprawl, as physical landscape and terrain can strongly define the type and pattern of crime. The connectivity and inaccessibility of locations, transport network, economic profile, and political scenario of an area governs crime. It may be home, workplace, or any other place which is in proximity to it. Crime patterns change over space and time in an area and hence it is of immense help to police by using GIS maps and spatial crime patterns to capture and study the patterns for better crime control. The ability to access and process information rapidly while displaying it in a spatial and visual medium permits agencies to distribute resources quickly and more effectively. GIS can be used as a tool to identify factors that are contributing to crime, by creating crime maps and providing solution to society through crime analysis by getting clusters and hot spots. GIS can aid huge amounts of location-based data from multiple fronts. It enables the user to layer the data and view the data most critical to the particular problem or operation.

The following case study of Ajmer City (Rajasthan, India) focuses on highlighting the usage of geospatial techniques in curbing crime occurrences in the region. Such investigative research interventions enable the city police towards enhanced spatial crime mapping and demarcation of hot spot zones towards better crime prevention control. This micro-level research aims to benefit and assist the police officials towards generating their surveillance plans and crime control strategies in the region. The basic objective of the study conducted was to identify crime-prone zones through crime mapping with probability of occurrence based on past incidence of various crime locations with changing urban profile of the city. This research helped to make available spatial crime hot spots so as police can plan better surveillance plans and scheduling so as to minimize crime incidences and take preventive action in short time based on minimum distance maps. Such investigative interventions provided a spatial decision-making system to Ajmer City police for better surveillance scheduling on a temporal basis. On the basis of acquired data, police officials trace local suspects and enhance patrolling in the affected areas.

The geospatial data, also known as geographically referenced data, defines both the location and the characteristics of the piece called spatial feature which may contain of roads, land extents, and vegetation on the earth surface (Burrough, 1986). In this study, the different types of crime such as robbery, home/shop breaking, automobile theft, kidnapping, rape, murder, and other theft have been considered. The crime incidences are recorded based on a GPS survey of the area. GIS open source and proprietary software are used for capturing, storing, converting data format, exploring, and displaying geospatial data.

GIS study the area through mapping liable to crime both spatial and statistical analyses using the appropriate tools such as neighborhood and correlation analysis, respectively. Geospatial technologies help to capture the spatial heterogeneity of the different types of criminal activities and security resources and thus to develop a spatial connection between the events in a specific region of attention. GIS helps the police to plan meritoriously for emergency response, decide mitigation priorities, analyze historical events, and forecast future events. One of the major activities that has to be performed by the crime investigation department is mitigation of hot spot locations where the number of crimes happening are more. Using hot spot techniques such as spatial analysis, interpolation, and spatial autocorrelation, the high concentration crime-occurring areas can be found. Such hot spots are small zones that have a great deal of crime or a disorder, even if there may be no common criminal; and crime analysis tries to link these to underlying social situations. Though there is no theoretical basis, hot spots are areas of imaginary boundary where there is recurrence of crime incidents.

6.3.1 Study Area: Ajmer City, Rajasthan (India)

Ajmer is considered the heart of the desert state Rajasthan of the Indian subcontinent. Ajmer is situated in the northwest region of India and is surrounded by the Aravalli Mountains. It lies on the lower slopes of the Taragarh Hills, and the massive rocks of Nagpahar separates Ajmer from the Thar Desert lying to the west. Ajmer geographically lies nearly in the center of Rajasthan (India) between 25°38' to 26°58' North latitudes and 75°54' to 75°22' East longitudes. Ajmer is blessed with a green oasis wrapped in the barren hills which has Tonk and Jaipur districts in the east, Nagaur in the north, Pali to the west, and Rajsamand and Bhilwara districts to the south. The city is at a distance of 135 km from the state capital, Jaipur, and 391 km from the national capital, New Delhi.

Ajmer City is a municipal corporation divided into 55 wards. The city is in the headquarters of the Ajmer district and is a popular religious and tourist destination. Its area is 219.36 square kilometers. The topographical map of Ajmer City (Fig. 6.1) gives a lucid presentation of the geographical details of the city region.

The geographical understanding of the region along with the demographic pattern helps in building better understanding of the crime arousal, crime trends, and pattern in a region. The population of Ajmer City is exceeding 0.5 million (2011) people, making it the fifth most populous city in the state of Rajasthan, India. The major demographic characteristics are given in Table 6.1. Ajmer City has witnessed a drastic change in terms of its urban profile and demographic transition in the past few years. Past researches have proven that crime occurrence is expected to have a strong relationship with demographic profile and occupational structure in any region.

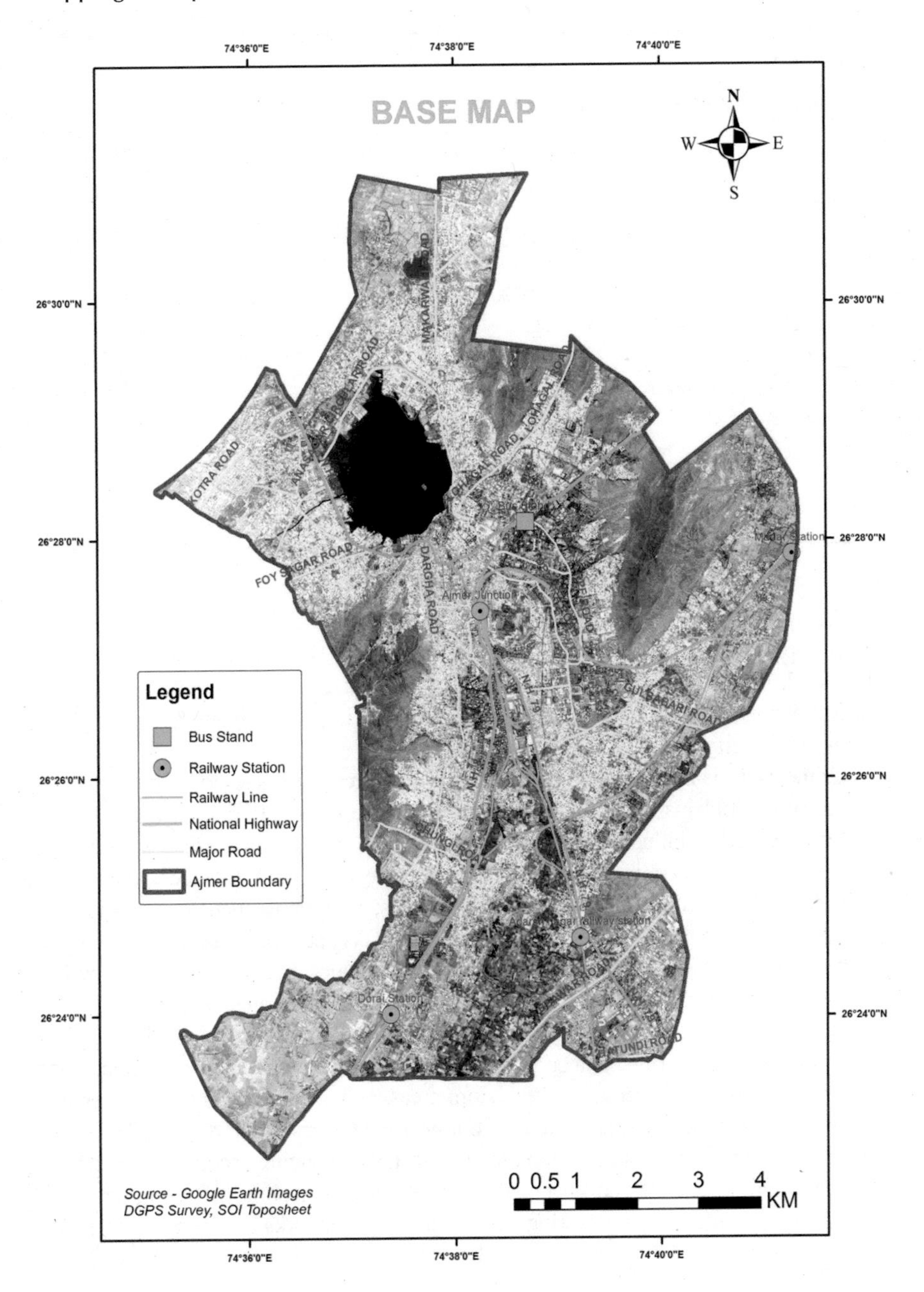

FIGURE 6.1 Base Map of Ajmer City, Rajasthan (India).

TABLE 6.1
Demographic Details of Ajmer City

Item	Persons	Males	Females	Rate/Percent
Population	542,321	278,545	263,776	947
Literacy	416,511	226,757	189,754	76.8%
Illiteracy	125,810	51,788	74,022	23.2%
Workers (Total)	174,922	143,668	31,254	32.2%
Cultivators	803	629	29	-
Agriculture laborers	896	733	163	-
Household industry workers	8,104	5,419	2,685	-
Other workers	152,076	127,707	24,369	-
Non-workers	367,399	134,877	232,522	67.8%

Source: Census Commissioner, India, 2011

The population density map (Fig. 6.2) highlights the innerwards areas with high population density in Ajmer City in the year 2011. The center city region is crowded, with areas along dense road networks, narrow lanes, and market complexes notified to be more densely populated in comparison to the peripheral areas of the city.

Crime analysis is the method used by the law enforcing agencies to lessen, avoid, and solve crime problems with criteria that determine the potential crime area for decision support. A strong influence of land use is limited to their immediate surroundings strongly attracting crime spots are alcohol outlets, cultural facilities, commercial buildings, bars, and low-income housing colonies; in contrast, depots-transports, gardens, and grandstands are strongly detracting. As per the master plan (Ajmer Development Authority) of the city, the maximum area of Ajmer City is under built up or in residential use. The comparative study of the Land Use Land Cover (LULC) maps strongly emphasizes (Fig. 6.3) the changing urban profile of the city from 2000 to 2017 and how it has transformed to a congested urban agglomeration with the development of narrow crowded lanes, dense markets, slum areas, and wasteland. These changes have deeply impacted the type and pattern of criminal incidences in the region. The city has undergone a complex urban evolution process with the changing occupational sector and emphasis on the tertiary and quaternary services than agriculture now. The urban advancements contributing to towering apartments, road networks becoming complex, and congestion in the old city center region has added to the arousal of various social and economic issues in the region.

The increasing red color and depleting yellow tint in the LULC of 2017, clearly earmarks the shooting up of the built-up area and change in occupational structure. The agricultural section of the society has now been involved in the other services. It is very essential to have a decadal comparison of the city growth for establishing any analysis form of crime existence.

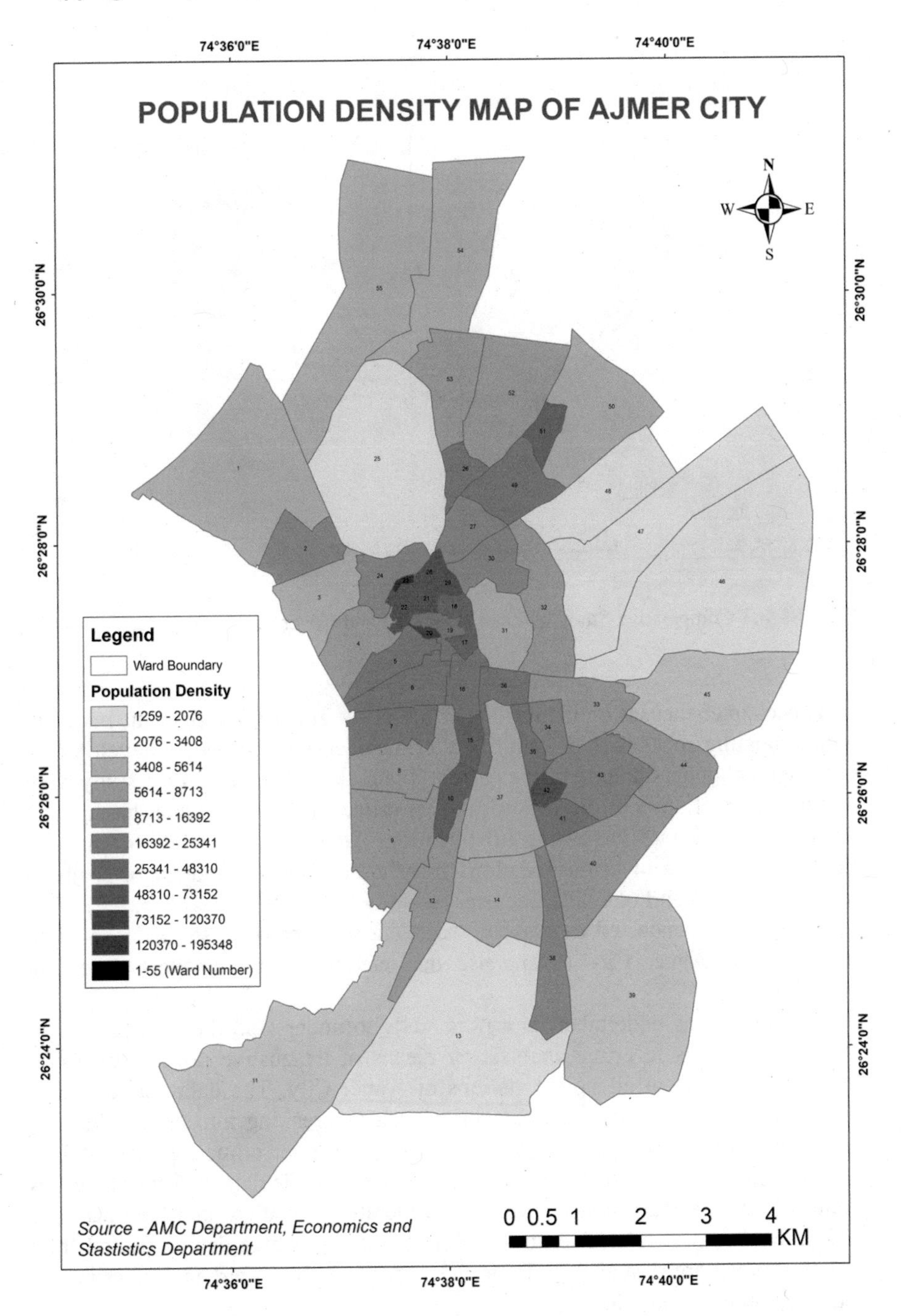

FIGURE 6.2 Population Density Map of Ajmer City, Rajasthan (India).

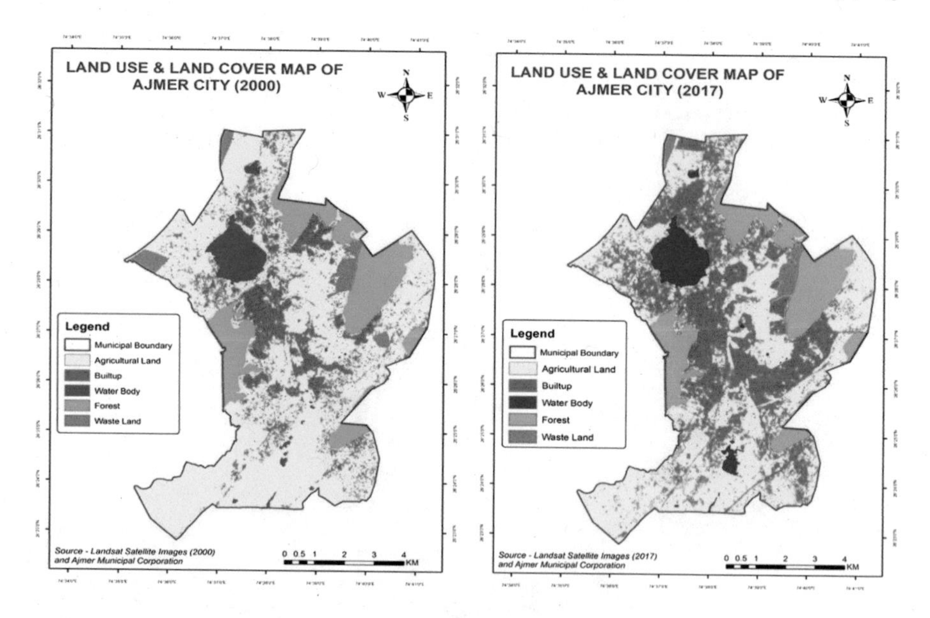

FIGURE 6.3 Comparative Study of Land Use and Land Cover of Ajmer City, 2000–2017.

This chapter focuses on identifying crime-prone zones through crime mapping with indicating zones with probability of crime occurrence based on developing a geographical understanding of the region. The major objectives of the study were identification of susceptible pickpockets of crime in the area which has greatly been influenced with increasing urbanization. Attempts were made to generate Euclidean distance of crime locations from police stations and major roads to understand the trend of crime occurrence along transport routes (Fig. 6.5). This study was also conducted to evaluate the relationship between sex ratio and rape incidences in Ajmer City to appraise the interrelationship amongst dependent variables.

This study was undertaken as a pilot study to understand the applicability of various GIS tools in crime analysis by means of exhaustive primary data collection surveys from all police stations of Ajmer City. The data was collected under the different recognized categories such as breaking into homes, kidnapping, murder, rape, automobile theft, robbery, and other crimes, etc., for all 55 wards from 2009 to 2014. The supporting socio economic data of Ajmer City was collected for sex-ratio, non-working and illiterate population, and land use land cover. The ward wise total crimes and non-spatial attribute data collected from police stations are provided in Table 6.3, where nearby suburban area crime incidences reported are 63.

The crimes in different police stations are listed in Table 6.2 and have been plotted in Fig. 6.4 within the police station boundaries in the city. It is important to understand that the police station boundaries include both the urban and rural profile of the city and are situated beyond the city limits.

TABLE 6.2
Ward-Wise Crime Incidence Record and Non-Spatial Data of Ajmer City

Ward No.	Total Crimes	Total Population	Illiteracy	Non-Working Population	Sex Ratio
01	10	18,719	4,293	12,512	972
02	4	11,110	2,005	7,647	963
03	27	12,902	2,814	8,652	997
04	21	12,492	5,608	9,103	896
05	11	8,177	2,776	5,396	962
06	7	11,422	2,841	7,585	989
07	0	9,190	2,780	5,927	946
08	5	11,859	3,769	7,984	964
09	4	9,475	2,310	6,233	978
10	3	8,927	3,641	5,819	953
11	5	16,493	3,359	11,076	941
12	1	9,345	1,479	6,495	967
13	8	15,746	4,650	11,157	956
14	3	5,802	1,056	3,964	922
15	1	6,497	1,374	4,468	982
16	7	7,261	1,177	4,823	955
17	0	6,773	1,446	4,564	954
18	0	5,048	730	3,335	947
19	0	6,885	1,362	4,607	924
20	0	4,959	1,389	3,739	969
21	0	5,998	1,010	4,052	968
22	3	6,380	1,723	4,284	898
23	0	11,621	4,134	8,141	957
24	43	10,141	3,713	6,767	944
25	74	9,758	2,019	6,621	955
26	2	9,743	1,659	6,550	990
27	4	5,479	782	3,792	992
28	1	6,312	1,318	4,270	963
29	0	6,051	1,052	4,004	960
30	1	6,376	1,091	4,331	946
31	4	6,784	1,435	4,670	905
32	0	8,885	1,604	5,985	989
33	1	5,807	1,101	3,704	953
34	0	9,479	2,068	5,694	1,133
35	0	9,516	1,826	6,112	958
36	1	5,497	895	3,669	902
37	4	8,919	1,420	6,279	923
38	0	10,057	2,355	6,882	948
39	5	9,128	3,461	6,279	935
40	0	12,121	2,482	8,382	951

(Continued)

TABLE 6.2 (Continued)

Ward No.	Total Crimes	Total Population	Illiteracy	Non-Working Population	Sex Ratio
41	0	10,382	1,578	7,285	946
42	0	9,295	1,470	6,366	978
43	0	9,861	1,743	6,733	962
44	0	11,918	1,990	7,934	973
45	0	12,901	3,178	8,703	830
46	6	19,069	4,161	13,105	958
47	6	7,067	1,213	5,014	952
48	0	13,392	3,109	8,410	711
49	1	8,199	1,278	5,414	960
50	2	13,486	3,260	9,552	961
51	1	11,505	3,232	8,111	940
52	3	9,793	1,891	6,611	964
53	2	9,909	1,886	6,549	968
54	0	14,860	2,140	10,006	966
55	11	17,550	5,674	12,052	923

The city has nine police stations (PS) and one all-female police station (Mahila). It has four major national highways, namely NH-8, NH-58, NH-79, and NH-89 passing through the city region. The road network has been taken from open street maps and the crime incidences were plotted to infer any trends in the two. The standard deviational ellipse was used to understand and study the potential distributional trend for a set of crimes and help identify its relationship to crime arousal.

The study focuses on analyzing the urban profile of the city and its trend of growth, keeping in mind the concept of geographical space, using ward wise maps from Ajmer Development Authority, Town Planning Department, and Survey Department. The shape files for police boundary and ward map were digitized for the recent years and the crime locations were plotted as points.

Several geospatial analysis tools were used to generate results in Ajmer City. As discussed earlier in Chapter 4, the kriging technique was applied for crime analysis in Ajmer City. It is a geostatistical interpolation technique that considers both the distance and the degree of variation between known data points while estimating values in unknown areas. It is a multistep process; it includes exploratory statistical analysis of the data, variogram modeling, creating the surface, and (optionally) exploring a variance surface.

Ordinary kriging was used (with local mean) on all crime locations within the police boundary on the source data set using a fishnet (Manfred and Arthur, 2010). The basic idea of using kriging was to predict the value of a function at a given point.

With the help of ordinary kriging, probability maps for crime incidences were generated based on incidents recorded, showing regions in the northwest and central city as more susceptible to crime incidences (Fig. 6.6). A gradual decrease

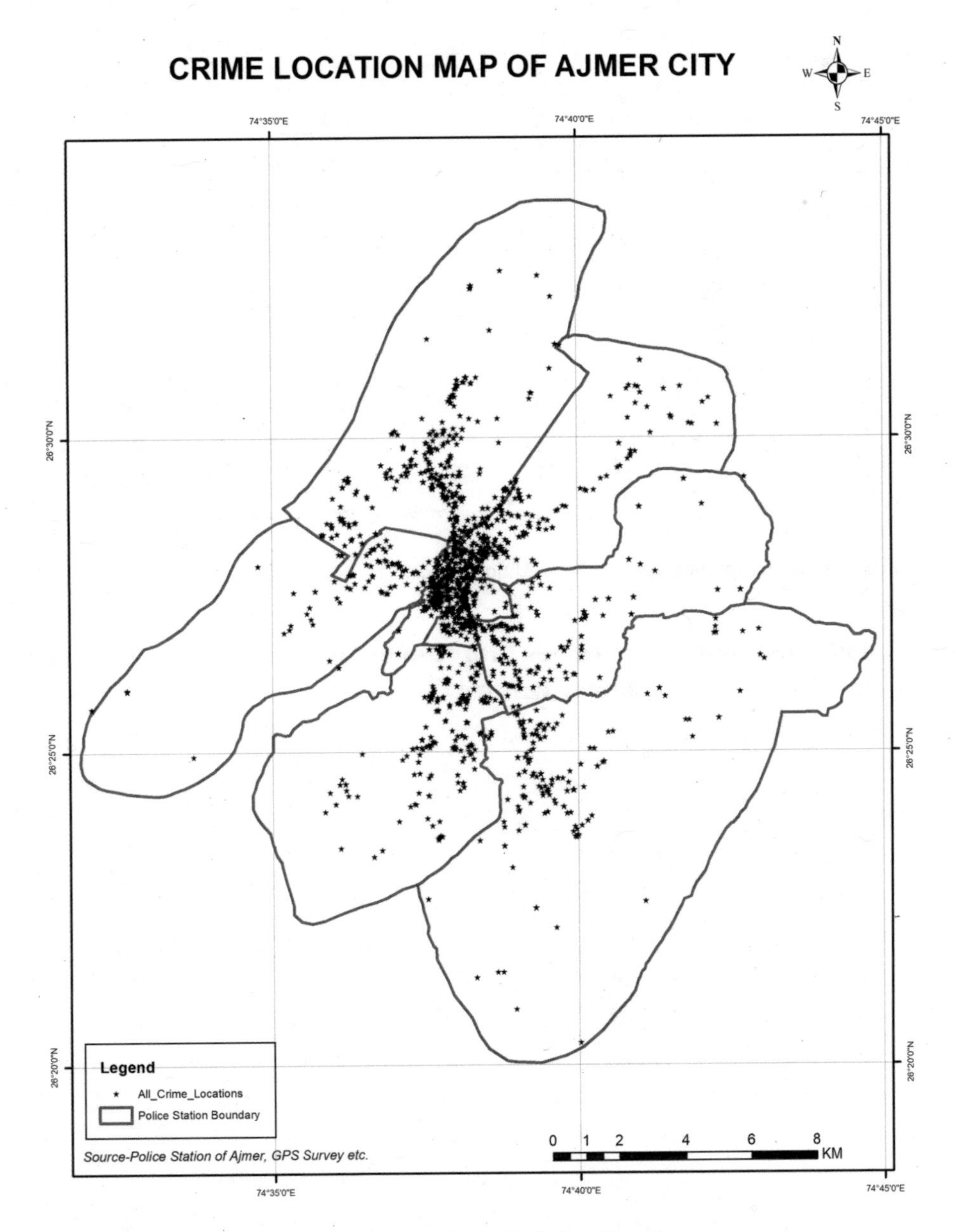

FIGURE 6.4 Crime Locations in Ajmer City with Police Boundary.

of probability of occurrence in the neighboring wards was observed with low values in the periphery region.

Further co-kriging technique was used which requires much more estimation, including estimating the correlation for each dependent variable with the independent variable. Co-kriging was used with crimes as the primary variable along,

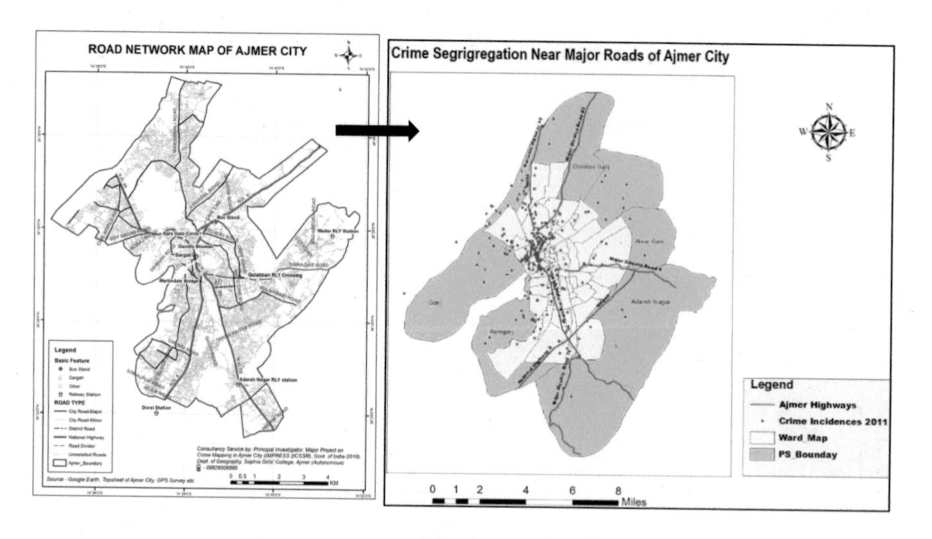

FIGURE 6.5 Crime Segregation Near Major Roads.

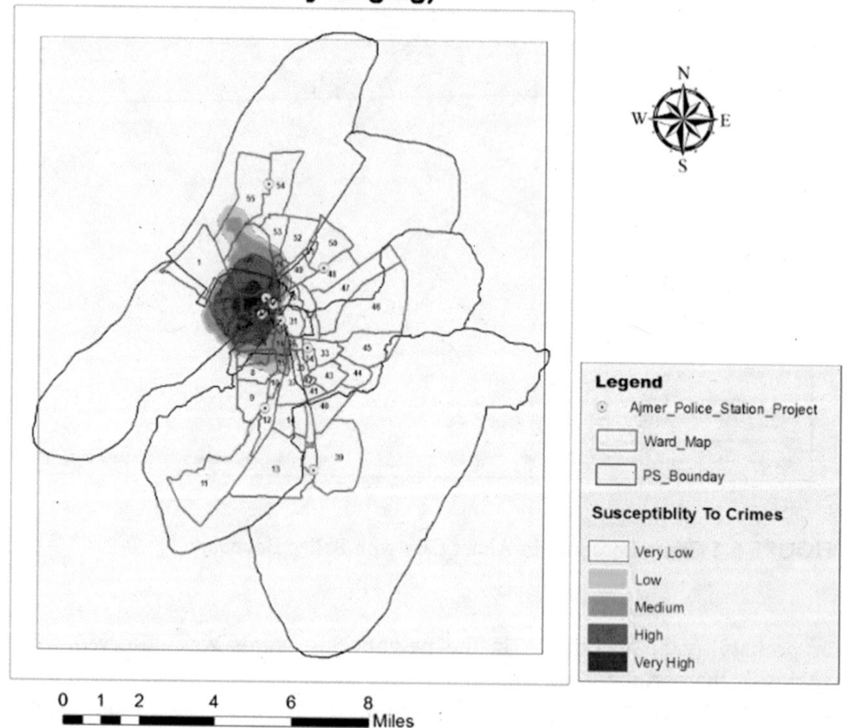

FIGURE 6.6 Crime Susceptibility—Ordinary Kriging Using Interpolation Technique.

TABLE 6.3
Correlation Coefficient Between Crime vs Non-Spatial Data

S. No.	Non-Spatial Attribute	Correlation Coefficient
1.	Illiteracy	0.991974628
2.	Non-working population	0.938512548
3.	Sex ratio	−0.279543878
4.	Population segregation	0.964581285

with the socioeconomic indicators with a strong correlation with the primary variable for further prediction.

The correlation coefficient of total crime with non-spatial attributes such as population, illiteracy, non-working population, and sex ratio were generated (as in Table 6.2) and except sex ratio all the other were found highly correlated. Here illiterate population, non-working population, and sex ratio were taken as the dependent variables and overall crime incidences for the year 2011 was taken as the independent variable. A very high positive correlation was observed between crime and the illiterate and non-working population, whereas a very high negative correlation was noticed between sex ratio and the registered rape incidences (as in Table 6.2).

The co-kriging technique results in formation of three high-crime-susceptible regions (Fig. 6.7) in the city center regions of Dargah, Ganj, Christian Ganj, and Clock Tower. The core city area has a religious hub with hundreds of pilgrims moving in and out of here, especially ward 55 in the north, 15 and 16 in the south-central, and the rest concentrated in 3, 5, 21, 22, 23, 24, and parts of 25, 26 wards. Medium crime susceptibility was observed mainly in wards 55, 2, 8, 10, 36, 17, 18, 29, and 27. The surrounding regions have low and very low susceptibility of crime (as in Fig. 6.7). In cases of rape incidences, maximum crimes were reported in the Alwar gate and Kotwali police stations, which lie in the least sex ratio area of the city. The co-kriging technique highlighted three crime-susceptible zones in the city situated in the center city region, northwest region, and southwest region along the major roads on the city. The results of interpolation technique states that the crimes are more susceptible in the northwestern and central part of the city in wards 2, 3, 5, 6, 20, 22, 24, and 25. There is a gradual decrease of probability of occurrence in the neighboring wards 15, 16, 26, and 55. It further becomes very low in the periphery region.

The Cluster Analysis Tool was used to identify spatial clusters of features with high or low values in the city with a given set of features (Input Feature Class) and an analysis field (Input Field). The tool also identifies spatial outliers. The Cluster and Outlier Analysis (Anselin Local Moran's) tool identifies concentrations of high values, concentrations of low values, and spatial outliers.

The Cluster/Outlier Type (Fig. 6.8) (COType) field distinguishes between a statistically significant crime cluster of high values (HH) as in wards of Ajmer City 2, 21, 22, 23, 24, 25, and 47; outlier in which a high value is surrounded primarily by low values (HL), and outlier in which a low value is surrounded primarily by

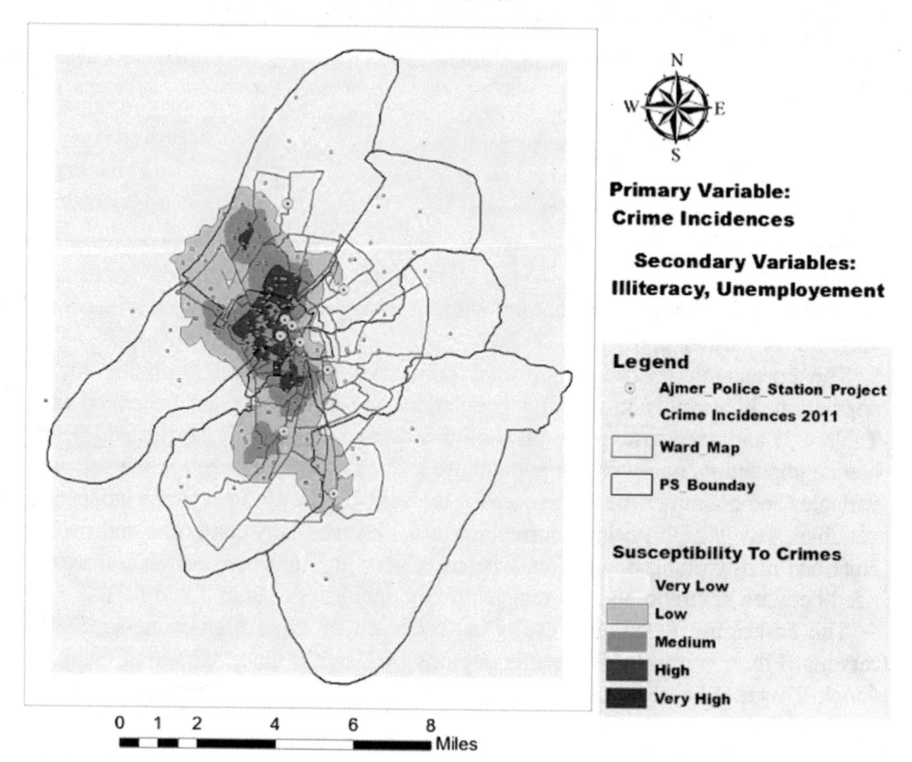

FIGURE 6.7 Crime Susceptibility—Co-kriging.

high values (LH) as in the case of ward 3 (as in Fig. 6.8). The surrounding region was observed insignificant to the occurrence of crime in this case.

The black check boxes indicate the high significant clusters of crime-prone regions in the city very similar to the results of spatial analysis using the interpolation technique. The white box indicates a cold spot with least crime occurrence. The different geospatial tools have been applied to yield results to test and verify both applicability as well as the utility of the same. The Euclidean distance tools describe each cell's relationship to a source or a set of sources based on the straight-line distance.

Often called the "as the crow flies" measurement, it's the shortest distance between two points on the map (Jill et al., 2001). In crime analysis, the most common methods of distance calculation are Euclidean and Manhattan. In Fig. 6.9, the maximum-crime spots are lying nearest to the minor and major roads in the Euclidean distance map. The densest network of roads in Ajmer City is in wards 3, 7, 8, 9, 15, 16, 22, 23, 25, 27, 28, 29, and 37.

The major concentration for crime with respect to distance is along the major district road 85. National highway numbers 89, 79, 58, 48, and 8 pass through the city and maximum concentration of crime is near NH 58 and NH 59. The major district roads more susceptible to crime are 79, 85, and 6.

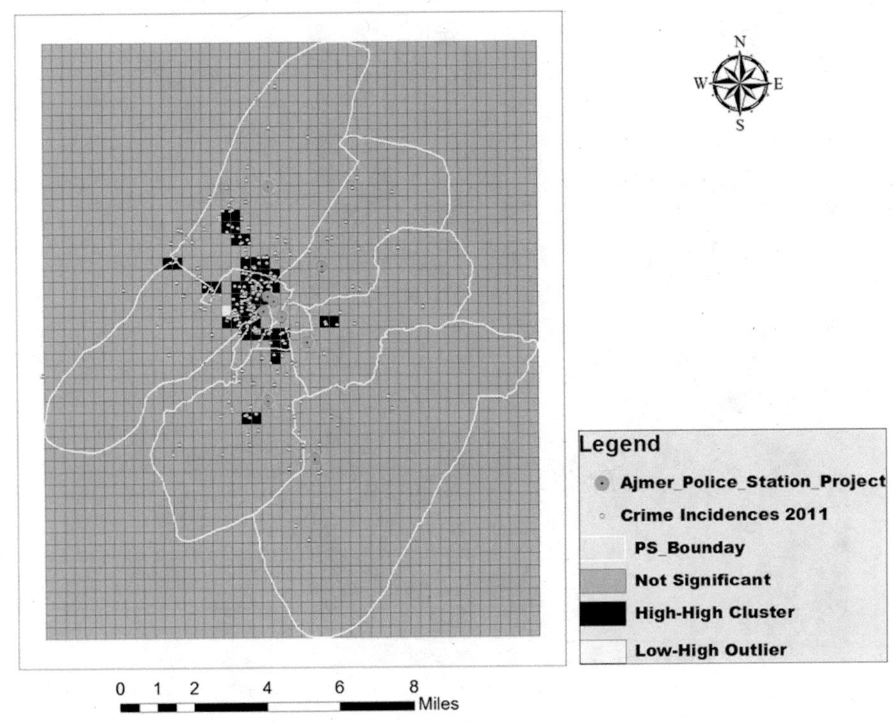

FIGURE 6.8 Cluster Analysis of Crime Using Anselin Local Moran.

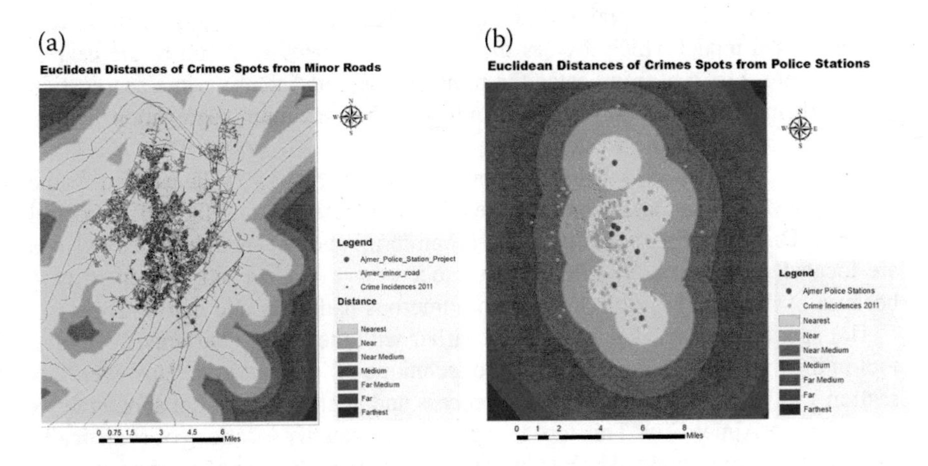

FIGURE 6.9 Euclidean Distances of Crimes from Minor Roads and Police Stations.

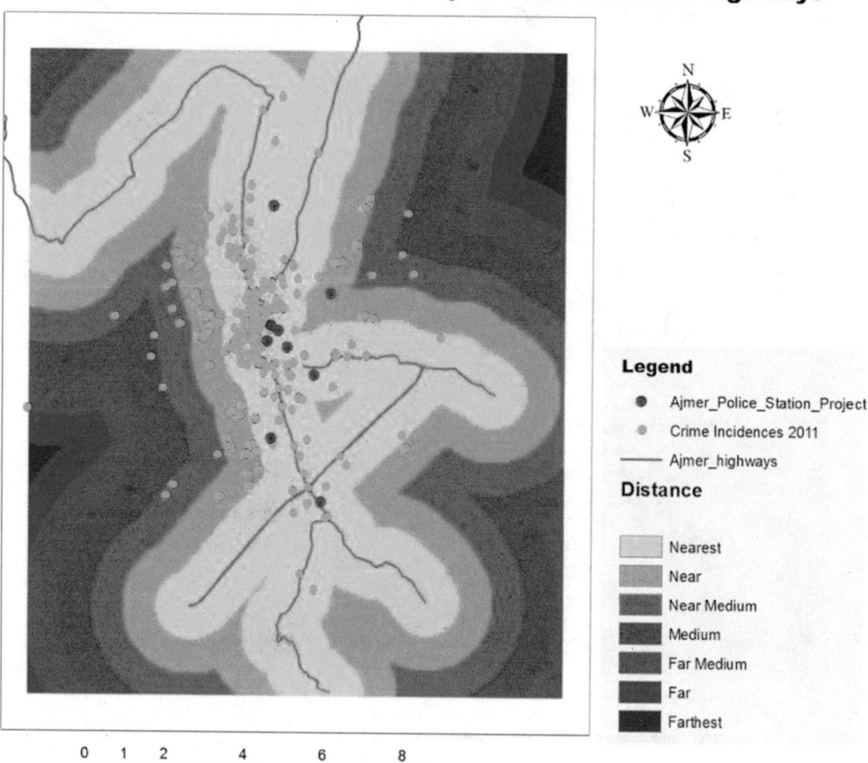

FIGURE 6.10 Euclidean Distance of Crimes from National Highways (NHs).

The Euclidian maps generated for minor roads, national highways, and police stations are basically the straight-line distances between two points on a plane. It is a raster-based model which divides the area and allocates each cell to the nearest input feature. Fig. 6.9(a) indicates the minor roads with intense crime segregation around them. The inner buffer 6.9(b) indicates the maximum segregation, with a trend pattern towards the western city.

This old city region is densely populated, has very narrow lanes, and is the overcrowded commercial zone of the area. It can be referred to as the CBD (Central Business District) of Ajmer City in the urban terminology. Though police stations are located in the vicinity, still the crime occurrence is frequently observed here because of narrow, dim light lanes with numerous pathways and market regions.

Hot spot analysis has been discussed earlier with the Inverse Distance Weighted tool in the initial chapters as an effective technique of interpretation. However, this section will help better understand the process and applications of hot spot analysis as applied in Ajmer City. This tool uses vectors to identify locations of statistically significant hot spots and cold spots in your data by aggregating points of occurrence into polygons or converging points that are in proximity to one another based on a calculated distance. The analysis groups feature similar high (hot) or low (cold)

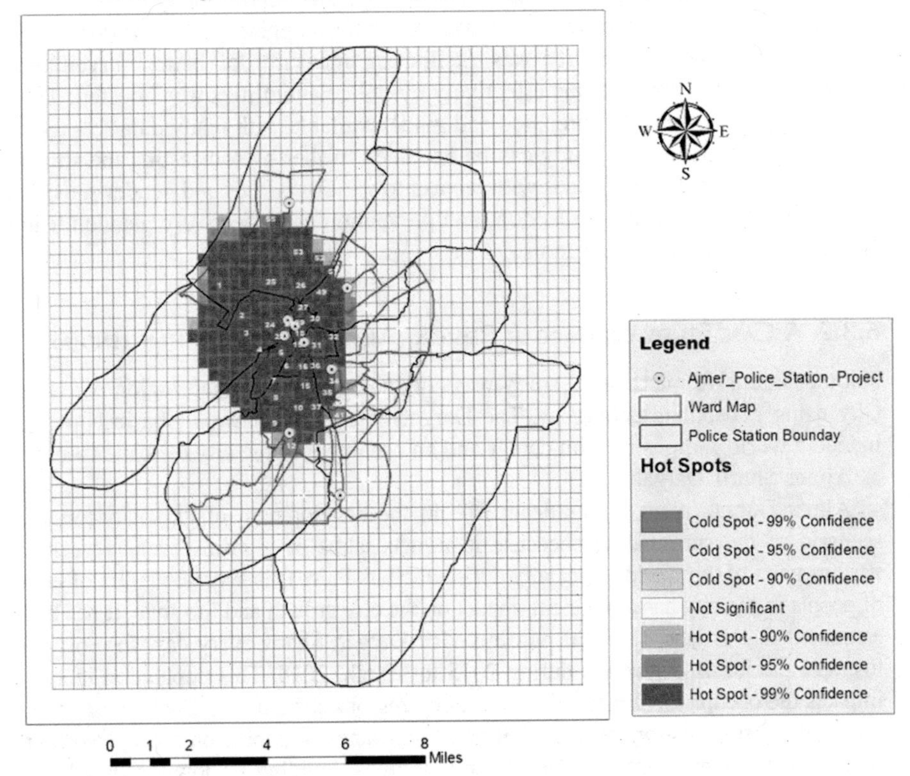

FIGURE 6.11 Hot Spots of Crimes in Ajmer City.

values in a cluster. The polygons usually represent administration boundaries or a custom grid structure.

The hot spot concentration, as shown in Fig. 6.11, is confined in the north western and central part of Ajmer City. It has higher values in the old city region with 99% confidence and has 95% and 90% confidence in the outskirts of the city.

The results of hot spot analysis were validated by conducting ground truthing surveys as well, which provided ground reality and unveiled the facts of the congested regions of Dargah, Ganj, Kotwali, Purani, Mandi, Madar Gate, Nala Bazar, and Kutchery Road areas having a very high intensity of crimes occurrence. The major reasons investigated behind such crime occurrences were the high non-working population, high illiteracy, people with mainly daily wage jobs, and street/footpath vendors of low per-capita. The other supporting field observation was the intake of locally made liquor and drugs.

A hot spot tool works by looking at each feature within the context of neighboring features. A feature with a high value is interesting but may not be a statistically significant hot spot. To be a statistically significant hot spot, a feature will have a high value and be surrounded by other features with high values as well. The local sum for a feature and its neighbors is compared proportionally to the sum of all features. One of the major objectives of crime mapping and identifying high crime-prone zones in the city was facilitating measures of better surveillance planning so as to control crime effectively by police authorities was boosted. Time scheduling maps were also prepared and made available with ready data sets to the policing departments. The study can be extended further by forecasting the shift in hot spots and crime-prone zones with respect to various crimes and gender sensitive crime investigation. Action, reaction, and prevention can effectively planned using GIS-enabled crime mapping tools.

6.3.2 A Case Study of Dargah Region, Ajmer City, Rajasthan (India)

Another intensive micro-level study was conducted in the Dargah region of Ajmer City using geospatial techniques. The Dargah Sharif of Khawaja Gharib Nawaj is indeed a world-famous pilgrim center and an ornament to the city of Ajmer, known as Ajmer Sharif Dargah.

Ajmer Dargah plays a significant role in the development of its socioeconomic scenario of the surroundings because it attracts lakhs of pilgrims every year from various parts of the world, especially during the Urs Fair. This sporadic influx of lakhs of people in the city has a pertinent impact on the city growth and life style. It has been estimated that around 1,50,000 pilgrims visit Dargah Sharif every day. Fig. 6.12 illustrates the influx of tourists during 2015 to March, 2018. This population increase impacts the occupational pattern of the nearby regions surrounding Dargah, especially during the festive season, by the addition of thousands of people on roads, leading to traffic jams, hiking of prices of basic commodities, building up low-fare residential complexes, and slums paving way to shortages of resources and crime increases.

The registered crime records were acquired from the Dargah Thana (police station) region for the years 2011–2018. It is essential to understand that to conduct any crime analysis, the prerequisite is substantial decadal real-time data,

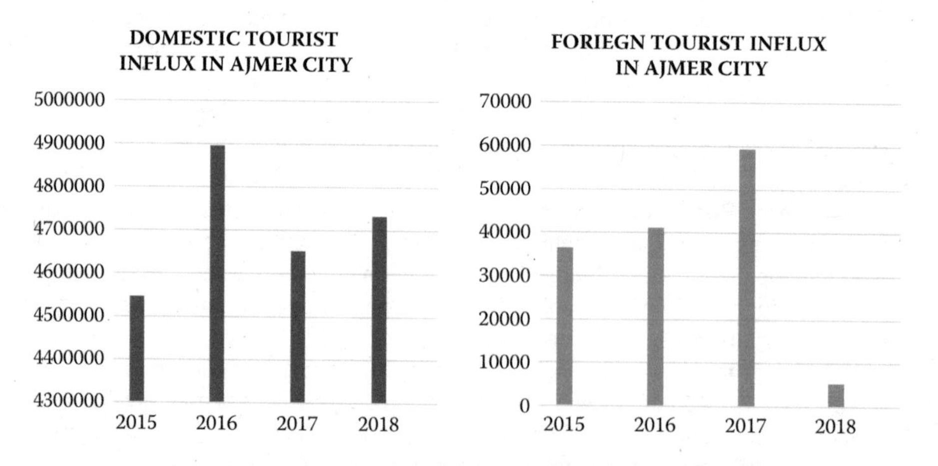

FIGURE 6.12 Tourist Influx in Ajmer City (2015–2018).

S.No	Crime Type	Date	Real Time	Location	LATITUDE	LONGITUDE	Specification
1	Molestation	12/07/2009	9:00:00 AM	Bhata Bhav Andarkot	26.454930°	74.626336°	
2	Molestation	16/07/2009	4:00:00 PM	Chodhar Mohalla Nala Bazar Dargah Police Station	26.457146°	74.629995°	
3	Molestation	27/08/2009	12:00:00 PM	Bhata Bhav Mandir Road Pani ki Tanki Dargah Police Station	26.458443°	74.628734°	
4	Molestation	21/09/2009	8:15:00 PM	Imli Mohalla Dargah	26.458093°	74.630217°	
5	Rape	03/05/2009	7:30:00 PM	Near Dargah Hotel Dargah Police Station	26.456864°	74.628770°	
6	Molestation	22/05/2010	10:15:00 AM	Saregram Mohalla Dargah	26.455305°	74.628596°	
7	Molestation	30/07/2010	6:30:00 PM	Dhan Mandi Dargah	26.453879°	74.635251°	
8	Kidnapping	08/11/2010	3:30:00 PM	Near Nijam Gate Dargah	26.457146°	74.628124°	
9	Rape	01/10/2010	11:00:00 AM	Near Dargah	26.457282°	74.627618°	
10	Rape	19/01/2011	5:00:00 PM	Imam Bada Khadim Mohalla	26.455184°	74.629815°	
11	Rape	15/06/2011	5:34:00 PM	Pannigram	26.457349°	74.627919°	
12	Rape	10/07/2011	4:36:00 PM	Garib Nawaz Colony Andarkot	26.453375°	74.621814°	
13	Molestation	16/02/2012	6:00:00 PM	Desh Bali Mohalla	26.482308°	74.649238°	
14	Kidnapping	21/04/2013	10:00:00 AM	Dargah Bazar	26.458346°	74.627911°	
15	Kidnapping	12/03/2013	9:00:00 PM	Taragarh Dargah	26.442245°	74.616916°	
16	Rape	02/01/2014	3:00:00 PM	Phool Babdi Andarkot	26.454261°	74.624546°	
17	Rape	27/01/2014	1:20:00 PM	Khadim Mohalla	26.455480°	74.627933°	
18	Rape	10/06/2014	3:50:00 PM	Dargah Police Station Parisar	26.458188°	74.628424°	
19	Rape	07/03/2014	4:15:00 PM	Muskan Guest House Dargah	26.458272°	74.627433°	
20	Kidnapping	21/01/2014	5:00:00 PM	Andarkot Dargah	26.454137°	74.624626°	
21	Kidnapping	31/08/2014	3:05:00 PM	Phool Babdi Kamla Andarkot	26.454282°	74.624700°	
22	Kidnapping	15/10/2014	7:30:00 AM	Gulab Kanwar School Dhan Mandi	26.453408°	74.634887°	
23	Kidnapping	06/04/2015	2:00:00 PM	Yaqoob Guest House Bhandara Gali	26.457940°	74.629778°	
24	Kidnapping	03/06/2015	1:05:00 PM	Jhalra Dargah Sharif	26.456038°	74.627785°	
25	Kidnapping	08/06/2015	2:00:00 PM	Lala Ji Ka Makan Lakhan Kotri	26.457518°	74.626255°	
26	Kidnapping	27/07/2015	11:35:00 AM	Dargah	26.455893°	74.628328°	
27	Rape	04/02/2015	1:00:00 PM	Mannat Guest House Imli Mohalla	26.457841°	74.630485°	
28	Rape	20/12/2015	8:15:00 AM	Pannigram Chock Dargah	26.457800°	74.628263°	

FIGURE 6.13 Record Sheets for Crime Records in Ajmer City.

which is a major constraint. As acquisition of such data is a herculean task, with approval and cooperation of the district administration, superintendent of police, and policing officers isn't that easy. Even if permission is granted, it becomes too tedious to sit for hours at police stations and keep noting down crime records for every single crime that has occurred from the official documentation data register (*Jurime*) of the police stations (Fig. 6.13). After crime data acquisition, to generate any substantial output or investigation with geospatial techniques, it becomes imperative to have at least decadal point data for all crimes in the region besides the geographical, social, and economic data sets as dependent variables. The data is initially notified in simple text format and later converted into excel sheets for GIS applications.

Fig. 6.14 discusses the rape incidences plotted in the Dargah Thana (police stations) boundary. The map clearly indicates that the concentration of crime is towards the northeastern direction (i.e., the city center around the Dargah region). Several maps were created using the acquired data, as per the research and investigation objective such as maps indicating the rape incidences in the region.

IDW is based on the assumption that the sampled points closer to the predicted points have more influence on predicted value than the sample points farther apart (Jill et al., 2001). Thus, IDW predicts the estimated value by averaging over all the known measurements, and assigning greater weight to nearer points. It can be interpreted with reasonable certainty that the darker tones in the IDW of Ajmer City (Fig. 6.15) represent the maximum values and Dargah region lies in the same. The major reasons responsible for accelerating crime incidences in the region are shortage of basic amenities such as sanitation, food, and shelter; this becomes a concern and leads to arousal of crime and other related social issues. Dargah Sharif Ajmer is a tag point for its adjacent areas as they comprise several market areas, low-fare hotels and restaurants, commuter zones, and complex transport networks being used by the mainly tourists (Jaayreen) visiting Khwaja Saheb Dargah Shareef.

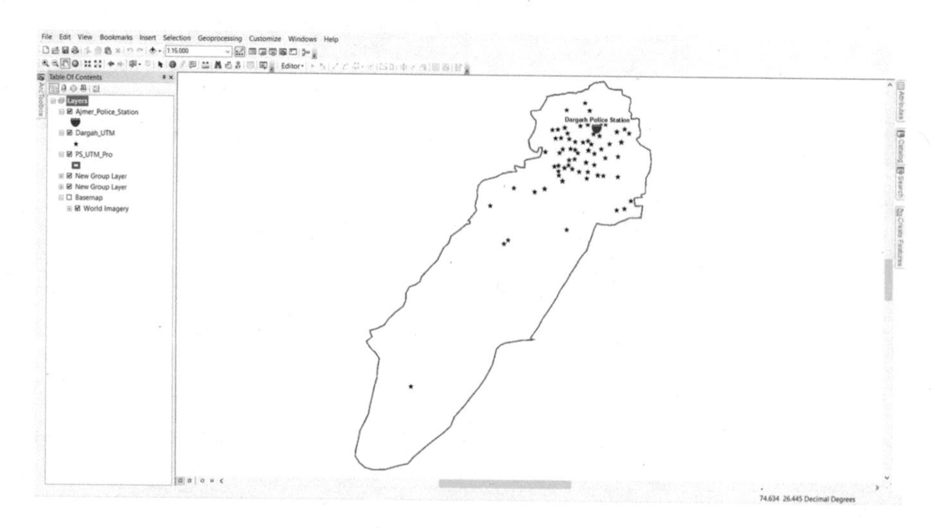

FIGURE 6.14 Rape Incidences in Dargah (Ajmer).

A hot spot analysis was conducted in the Dargah region of Ajmer city to demarcate the pickpockets of crime susceptibility against women in the region. The major cases registered here were under sections 376-IPC, 406-PC, and 498-A, which cover rape, dowry, and domestic violence.

The Dargah region is registered to have fast-tracking record of crime against women. The red-colored regions in Fig. 6.16 indicates the areas of violence against women in the Dargah region, mainly the Khadim Mohalla, Ander Kot, and Dargah Bazaar associated region of such dire conditions. Another crucial reason for the

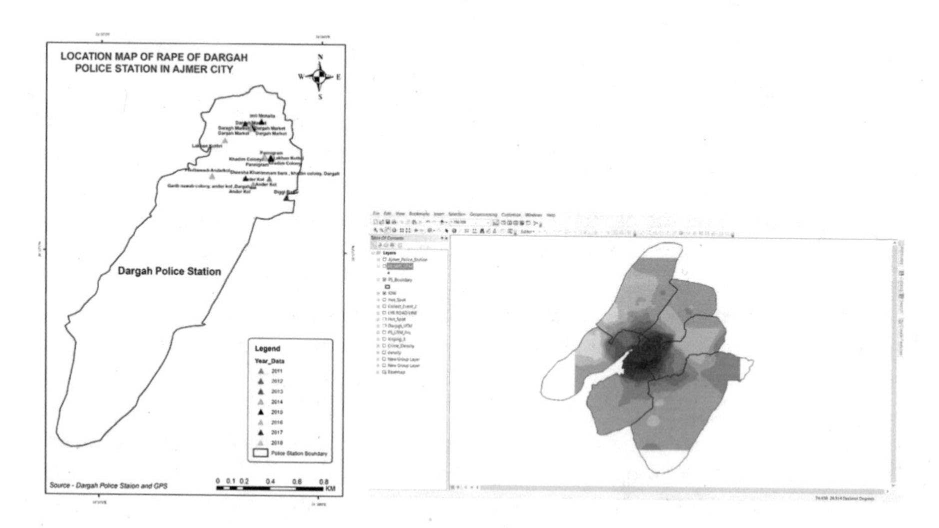

FIGURE 6.15 Rape Incidences Registered in Dargah Police Station Projected from the IDW Map of Ajmer City.

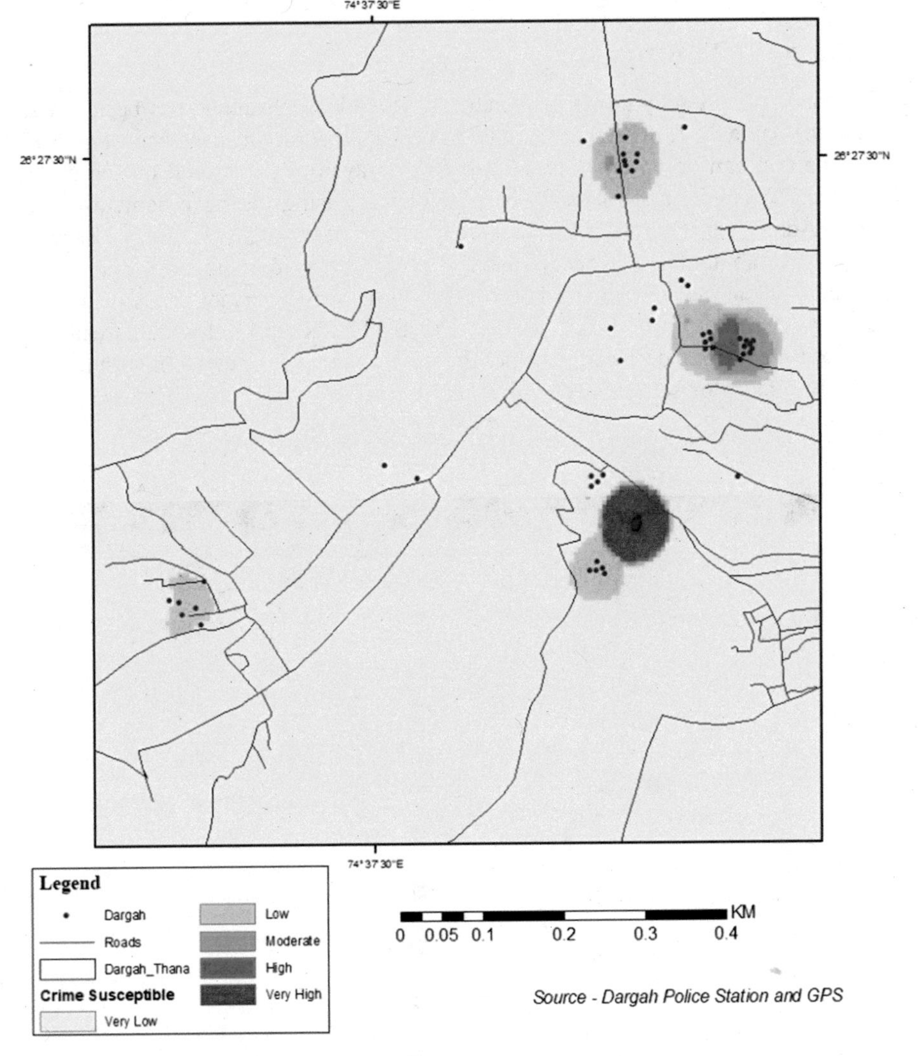

FIGURE 6.16 Crime Susceptibility Against Women in Dargah Region, Ajmer City.

hike in crime incidences in this region is its strong transport connectivity to all major roads of the city, which provides an easy escape for culprits. It is astonishing to mention that the apathy of crime analysis studies, especially in India, is that nearly 50% of the data of crimes occurred against women remains unregistered. Either due to family pressure, religious constraints, gender bias, or illiteracy, many females in the country still prefer to refrain themselves from getting justice and don't file their cases in the government police stations. With change in the education scenario and government interventions, a gradual increase in registering of cases can be seen, but much is yet to be done.

6.3.3 SATARK Application

To combat day-to-day crime registration, a "Safety Application" has been devised with the name SATARK in Ajmer City to provide safe city management (Fig. 6.17).

The app can be easily downloaded from Playstore and stored on the mobile device. The registration process is quite simple, using the governement-authorized Adhar card number (citizenship card). The app displays necessary information about the various registered crime-reported incidences and creates decadal crime incidence records for all minor as well as major roads of the city, clearly visible on the user's screen. The SATARK app helps demarcate safe routes to reach any destination as per the user's choice and selection with the help of different color schemes (Fig. 6.18).

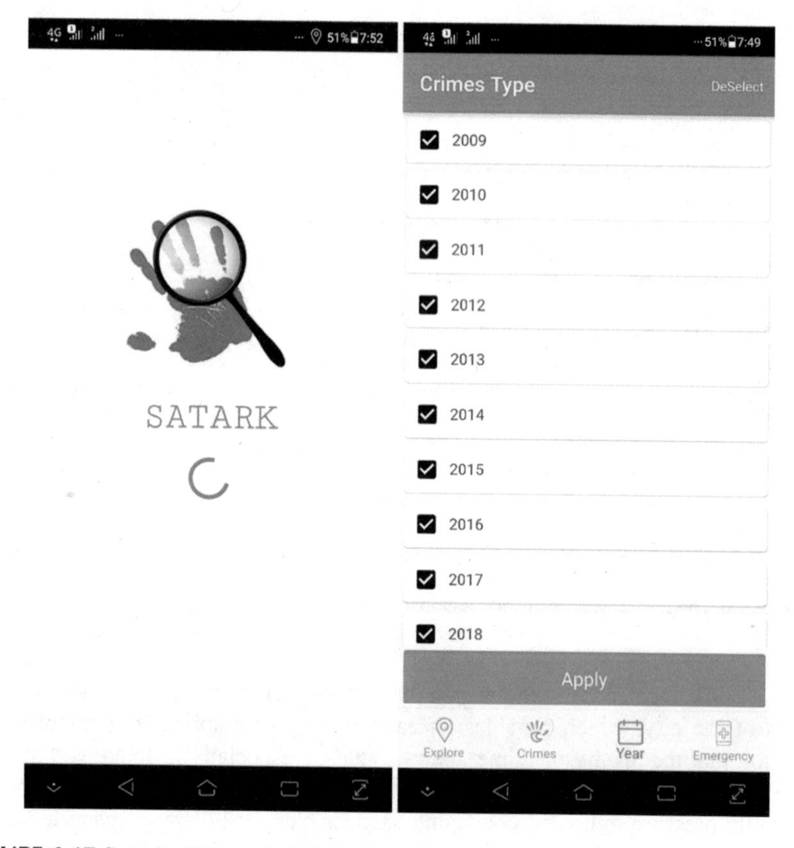

FIGURE 6.17 Security Features of SATARK Application.

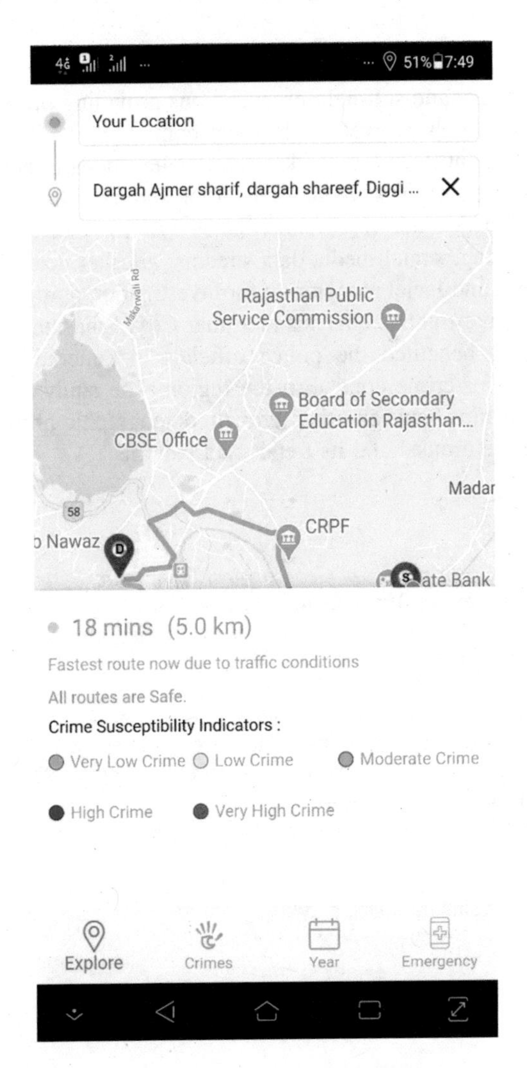

FIGURE 6.18 Navigational Features of SATARK Application.

The app has the provision of a panic button also, which when pressed sends the immediate location of the victim to his/her family, friends, or relatives based on the prior listed numbers, to seek assistance. The app not only provides safe transportation and movement in the city, but also contributes in the creation of unregistered crime data sets for further research and investigation.

Concluding Remarks

There has been a growth in social concerns about crime and fear of crime such that law and order has become a major public policy issue today. Besides the proactive paradigm shift and efficacy of the police department, numerous private security services are also developing rapidly. Recent years have seen the emergence of operational voluntary and self-help organizations providing protection and security measures to individuals as well as to local communities. Geospatial techniques have immensely contributed in making the police department robust and self-sustainable. The local crime statistics acquisition and monitoring stations contribute in data acquisition through video surveillance systems; license plate readers; vehicular GPS tracking; social media data streams; gunshot detection, radiation, and chemical sensory; and facial recognition for investigation purposes. Crime mapping and micro-level investigative analysis in Ajmer City, Rajasthan (India), using geospatial techniques benefited the police officials in generating surveillance and scheduling plans for crime control in the region. The study revealed astonishing facts and relationships between urban growth, demographic profiling, and religious tourism with crime arousal and its trends and patterns.

REFERENCES

Burrough, P. A. (1986) *Principles of Geographical Information Systems for Land Resources Assessment*. Clarendon Press, Oxford, p. 154.

Hoover, L. (2014) *Police Crime Control Strategies*. Delmar Cengage Learning, US, p. 112.

Jill, M., Steve, K. J., and Kevin, J. (2001) *Using ArcGIS Spatial Analyst: GIS by ESRI*. Redlands, CA, USA, p. 92.

Manfred, M. F. and Arthur, G. (2010) *Handbook of Applied Spatial Analysis: Software Tools, Methods and Applications*. Springer, Heidelberg, Germany, p. 29.

Palmiotto, M. (2000) *Policing Corruption: International Perspectives*. Lexington Books, Maryland, p. 217.

Philip, G. M., John, H. M., and Timothy, A. R. (2000) *Analyzing Crime Patterns*. Sage Publications, London, p. 57.

The Kansas City Preventive Patrol Experiment. (1974) A Summary Report, Police Foundation, Washington, DC, p. 34.

7 Crime Monitoring and Management

Crime is a human phenomenon; therefore, its distribution in space is not random (National Institute of Justice). Prior to the use of computers, the law enforcing authorities had a tough job controlling crime; they had very limited resources compared to resources that modern crime fighters have today. Computers have really changed the dynamics of crime fighting; crime fighters are now armed with the latest technology, which helps them nab the culprits in no time. Crime analysis methods are important because they help to identify the different geographic patterns in criminal behavior. There are many conventional tools that allow crime data analysis, but GIS software can create a single visual output that combines multiple data layers into a meaningful output. The analysis between crime and other factors, for example, demography, housing, income, or social conditions, can lead to the understanding of the place and crime relationship based on the conditions compared. Traditional law enforcement approaches included confidential information through informers, street investigations, and undercover operations were effective ways of data collection.

However, data collection without data analysis is of no use, and the GIS allow effective integration and analysis of data to identify, apprehend, and prosecute suspects; it also helps the law enforcing agency to fight crime through effective allocation of resources. Today, with the rapid advancement of technology, a computer-based technique for exploring, visualizing, and examining the occurrences of criminal activity is essential. One of the more influential tools facilitating exploration of the spatial distribution of crime is GIS. The fundamental strength of GIS over traditional crime analytical tools and methods is the ability to visualize, analyze, and explain the criminal activity in a spatial context. GIS has evolved as a powerful analytical tool in the last few decades. GIS started in the era of the mainframe computer. At that time, it was a costly affair and not many law enforcing agencies could afford this technology, but the migration of GIS from mainframe to desktop computer has provided the law enforcing agencies a cost-effective option for crime control. Due to the decrease in prices of GIS software, this technology is now easily available to any law enforcing agency in the world. Information management has always been a main concern for law enforcement authorities, especially the location information. Certain environmental factors, such as the physical layout of the area, proximity to various services, and land use are likely to influence criminal behavior and it is necessary to take them into account when analyzing the crime data. The majority

(91%) of law enforcement departments using GIS software reported the use of geo-coding. This method is the initial and vital step in the creation of a geographical data collection database, so that the data includes exact time and location information (Markus and Mitasova, 2002). Consequently, modern GIS software allows law enforcement agencies to produce more versatile electronic maps by combining their crime databases of reported crime locations with digitized maps of the target areas.

7.1 ROLE OF NATIONAL CRIME RECORD BUREAU (NCRB)

NCRB was set up in India in 1986 to function as a repository of information on crime and criminals so as to assist the investigators in linking crime to the perpetrators. It was set up based on the recommendation of the task force and National Police Commission by merging the Directorate of Coordination and Police Computer (DCPC), Statistical Branch of BPR&D, Inter State Criminals Data Branch of CBI, and Central Finger Print Bureau of CBI (Saxena, 1997).

Mission: NCRB envisions to empower the Indian police with information technology and criminal intelligence to enable them to uphold the law and protect people and to provide leadership and excellence in crime analysis, particularly for serious and organized crime.

Objectives: The major objectives and focus areas of NCRB are as follows:

1. Create and maintain secure sharable national databases on crimes and criminals for law enforcement agencies and promote their use for public service delivery.
2. Collect and process crime statistics at the national level and create a clearing house of information on crime and criminals both at the national and international levels.
3. Lead and coordinate development of IT applications and create an enabling IT environment for police organizations.
4. Collate a national repository of fingerprints of all criminals.
5. Evaluate, modernize, and promote automation in the state crime records bureau and state fingerprint bureau.
6. Enhance training and capacity building amongst police forces in information technology and fingerprint science.

7.1.1 Major Components of NCRB

NCRB has an effective management system, which works under the following parts.

7.1.1.1 Crime and Criminal Tracking Network and Systems (CCTNS)

The Crime and Criminal Information System (CCIS) was implemented at the district level during the period 1995–2004. The Common Integrated Police Application (CIPA) was implemented at the police station level during the period 2004–2009 in three phases. The four major objective of this body are:

1. Creating state- and central-level databases on crime and criminals.
2. Enable easy sharing of real-time information/intelligence across police stations, districts, and states.
3. Improved investigation and crime prevention.
4. Improved service delivery to the public/stakeholders through citizen portals.

7.1.1.2 Citizen Portals

There are various services, as enlisted below, being provided in the citizen portals for the Indian national, such as:

1. Filing of complaints.
2. Obtaining the copies of FIRs.
3. Obtaining the status of the complaint/FIR.
4. Details of arrested persons/wanted criminals and their illegal activities.
5. Details of missing/kidnapped persons and matching them with arrested, unidentified persons and dead bodies.
6. Details of stolen/recovered vehicles, arms, and other properties.
7. Verification requests for servants, employment, passport, senior citizen registrations, etc.
8. Portal for sharing information and enabling citizens to down load required forms.
9. Requests for issue/renewal of various NOCs/permits/clearances.

7.1.1.3 Interoperable Criminal Justice System (ICJS)

ICJS has been mandated for integrating CCTNS, police with e-courts, e-prisons, forensics, fingerprint bureau, and prosecution, which are the key components of the criminal justice system. Implementation of ICJS ensures quick data transfer among different pillars of criminal justice system.

7.1.1.4 Crime Records Matching

The main objective of the crime record branch is to collect, collate, and disseminate information on crime, criminals, persons, and property for matching purposes. The branch utilizes the following software systems:

1. Vahan Samanvay – An online motor vehicle coordination system for the coordination of stolen and recovered motor vehicles across the country. Police, RTOs, and insurance sectors are the main stakeholders. The general public also benefits from this system.
2. Talash Information System – This system is used to maintain and coordinate information on missing, traced, unidentified persons and unidentified dead bodies.
3. Fake Indian Currency Notes System (FICN) – It is an online system for compilation of fake Indian currency data. Police, banks, investigating

agencies, and other intelligence agencies and ministries are stakeholders of this system.

4. Fire Arms Coordination System – This system is used for the coordination of missing/stolen and recovered firearms.
5. Color Portrait Building System – This system is used to create portraits of suspects based on the description given by victims and eyewitnesses.

   Since 2011, 600 persons (alive and dead) have been united with their families by matching photographs and other physical features.

   Since the launch of the online application Vahan Samanvay in 2014, 68,264 data has been captured and 30,577 stolen vehicles have been matched from different states.

7.1.1.5 Central Finger Print Bureau

The Central Finger Print Bureau (CFPB) came into existence in the year 1955 at Kolkata, India. The CFPB is an apex body in the country to coordinate, guide, monitor, and provide technical support to the state fingerprint bureau, as well as investigating agencies and international organizations such as INTERPOL, in all matters related to the fingerprint science. It maintains the records of 10-digit fingerprint slips of convicted persons and those belonging to specified categories (repository of over 1 million digitized 10-digit fingerprint slips). It is a computerized system of matching fingerprints on the basis of ridge characteristics. CFPB has done pioneering work in automation of fingerprints at the national level through the Automated Finger Print Identification System (AFIS). The software was jointly developed by NCRB and CMC Ltd. The similar kind of AFIS is now extensively being used by many states/UTs for digitization of fingerprints. CFPB also receives a number of finger print document cases, from different government departments, courts of law, banks, post offices, investigation agencies, etc. The other important tasks handled by CFPB are as follows:

1. It imparts training to police and non-police personnel, including an 18-week proficiency course in fingerprint science twice a year.
2. It is also mandated to conduct annual All India Board Examination (AIBE) for accreditation of the fingerprint experts of India.
3. Every year, CFPB also participates in the All India Police Duty Meet (AIPDM) for conducting Finger Print Test.
4. It compiles data collected from Finger States/UT and publishes annually-fingerprint tests in India.

7.1.1.6 Crime Statistics

NCRB brings out three annual reports (i.e., Crime in India, Accidental Deaths and Suicides in India, and Prison Statistics India). These reports are principal reference points for police officers, researchers, and media and policy makers. Besides, the

Bureau is also collecting crime statistics and anti-human trafficking statistics on monthly basis. After extensive and exhaustive deliberation with various stakeholders, the Performa for Crime in India, Monthly Crime Statistics and Accidental Deaths & Suicides in India was revised in 2014. Performa for Prison Statistics is under revision. Provisions for inclusion of seizures by Central Paramilitary Forces have also been made. NCRB has developed application software for Crime in India (CII), Monthly Crime Statistics (MCS), Accidental Suicide in India (ADSI), and Prison Statistics of India (PSI). The Bureau is also conducting Training of Trainers (ToT) on Crime in India and Accidental Deaths & Suicides in India and Prison Statistics India for officials of SCRB and prison departments of states/UTs. NCRB has been given the "Digital India Awards 2016" in the open data championship category with silver on 9th December, 2016, for updating more than 3,000 data sets on the Open Government Data (OGD) Platform India in open-source format. NCRB has digitized "Crime in India" since 1967 and "Accidental Deaths & Suicides in India" since 1998.

7.2 SAFE CITY CONCEPTS

A perspective public safety has emerged as an important function for governments across the world. It refers to the duty and function of the state to ensure the safety of its citizens, organizations, and institutions against threats to their well-being as well as the traditional functions of law and order. With more than half the global population today living in urban areas, a safe city is increasingly being considered essential in ensuring secure living and prosperity. Crime, violence, and fear in cities pose significant challenges. The basic principles of good governance must find a direct application in any urban safety strategy, aimed at reducing and preventing common problems of crime and insecurity. The United Nations, through its Habitat Agenda on Human Settlements (Habitat II, 1996), which was adopted at the Istanbul Conference, initiated a series of approaches and strategies to effectively reduce and eradicate violence and crime within the cities. The aim of the UN-Habitat Safer Cities program is to reinforce personal safety and reduce fear by improving safety services and accountability to the community (UNHSP Global Report on Human Settlements, 2007).

The Safer Cities program has the following building blocks.

7.2.1 Building Urban Safety Through Urban Vulnerabilities Reduction

The UN Safer Cities program defines vulnerability as the probability of an individual, a household, or a community falling below a minimum level of welfare (e.g., poverty line) or the probability of suffering physical and socio-economic consequences (homeless or physical injury) as a result of risky events and processes (as forced eviction, crime, or flood). Paying special attention to urban vulnerabilities and violence shall reduce the probability of crime and ensure a secure and safe city environment.

7.2.2 Building Urban Safety Through Urban Planning, Management, and Governance

Sustainable urbanization by emphasizing inclusive and participatory urban planning and local development practices, incorporates policy making and strategy development. This in turn promotes institutional and organizational development, resource planning, and management in order to enhance efficiency in governance.

7.2.3 Improving the Governance of Safety

Enhancing urban safety and social cohesion are issues of good urban governance. They intend to create a city where safety is improved for its citizens and neighborhoods, where there is fearless interactions among people and groups. These are prudent aspects of good governance which create an enabling environment for the inhabitants of the city, allowing improved quality of life and fostering economic development.

7.2.4 Safe City Concept: An India Perspective

The Constitution of India enjoins the Union to protect every state against external hostility and internal disorder in order to ensure that the governance of every state is carried out in accordance with the provisions of the Constitution. In pursuance of these obligations, the Ministry of Home Affairs (MHA) continuously assesses and monitors the internal security situation, issues appropriate advisories, shares intelligence inputs, extends manpower and financial support, and offers guidance and expertise to the state governments for maintenance of security.

As per the Seventh Schedule to the Constitution of India, police and public order are state subjects and, therefore, the state governments are primarily responsible for the prevention, registration, detection, and investigation of crime and prosecution of the perpetrators of crime within their jurisdiction (Parliamentary Debates: Official Report, 2006). However, the MHA supplements the efforts of the state governments by providing them financial assistance for the modernization of the state police forces in terms of weaponry, communication, equipment, mobility, training, and other infrastructure. The MHA from time to time introduces key initiatives to ensure that the states and the state security agencies are equipped with the latest technologies and systems to ward off any security threat to its citizens.

7.2.5 Elements of Safe City

An integrated surveillance system provides a safe and secure environment for the development and growth of these aspects. Integrated city surveillance in India's vibrant democracy and healthy amalgamation of the socioeconomic and cultural diversity is reflected in its cities. Schools, colleges, educational institutions, and technical and infrastructural resources such as airports, railway stations, power plants, refineries (including heritage and cultural buildings), and most importantly, its citizens are the source of a city's energy. There is a growing recognition of the interconnectedness and the interaction between factors that spark and drive crises and the ecosystem of security management that handles these factors. Furthermore, there is greater focus on the high economic and social costs because of the lack of a robust security management methodology.

The key components available in a Smart City concept in India are illustrated in Fig. 7.1. The major four entities that collectively form a part for a smart public safety and security work are:

- Detection and Integration: Cameras and sensors, CCTV cameras, ANPR, gunshot detection, sensors
- Analysis: Real-time alerts, analyze and index, searchable video, event search, pattern analysis
- Decision: Coordination center, dashboards, command center, incident management, investigative analysis
- Action: First responders, on ground action, police or medical, riot control, bomb disposal

Workflow for a smart public safety and security includes,

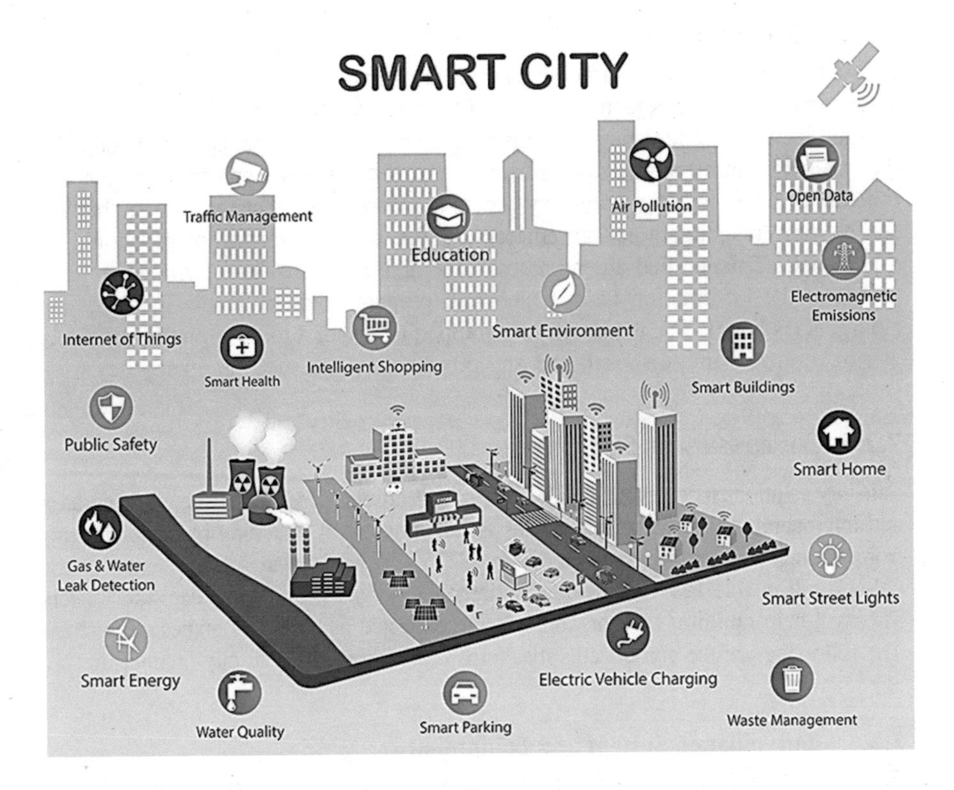

FIGURE 7.1 Key Components of a Smart City.
Source: Meungo, Glorieuxlaan, Vught, The Netherlands.

7.2.5.1 Detection and Integration

Capturing and collating data in real time is the lifeline of a safe city project, especially if the data pertains to an incident leading to a security threat. The idea is to collect data from a wide range of different sources in digital formats so that it can be efficiently processed and used in real time. The detection system may be driven electronically using high-definition CCTV cameras, Infrared Radars (IRs), lasers, handheld devices, or simply an incident reported by a vigilant citizen through a mobile or social media. It is imperative that the relevant data is obtained for recording and analysis. This information is then integrated from disparate sources and systems to create a dependable information base.

7.2.5.2 Analysis

It is essential to integrate the registered information and make it available to the security agencies in usable formats; the analysis tools help assist the agencies in the analysis and retrieval of crucial information from the captured data for pattern analysis, event search, and video retrievals. Information once retrieved in the desired formats helps in proactive planning and decision making. The dashboard reporting feature allows agencies to use the available information better and improve the strategic and tactical decision-making including anticipation, prevention, and resource deployment.

7.2.5.3 Decision

By establishing state-of-the-art command, control, and coordination centers, equipped to accommodate decision makers and facilitate decision making, dashboard reporting, incident management, and investigative analysis of the data collected using the real-time information, the infrastructure established enables effective and efficient decision making, with easy access to information. Action: A team that will act as first responders to an emergency for disaster management should be established. This unified threat assessment and response team will have ready access to the information base, which will prepare them to handle incidents and emergencies better. The team will be equipped to respond to on-ground action such as riot control, bomb disposal, and any other police and medical emergency.

7.2.6 Components of Safe City

Safe city solutions incorporate a wide array of technology-driven subsystems. Integration and interoperability of these subsystems is fundamental in obtaining better intelligence from various sources and sensors. From CCTVs to crisis management centers, technology will enable law enforcement, emergency services, and local decision makers. This will help optimize their response to the expected as well as unexpected mishaps. The following are the components that form the basis of the safe city architecture:

7.2.6.1 Surveillance System and Equipment

The focus of any safe city program is to provide officers and first responders with a shared security presence and an enhanced awareness through a system equipped with video surveillance cameras. The network of cameras collects data in the form of images or videos that are required to detect risks and respond to emergency situations. The CCTV camera technology has evolved over time, starting out as 100% analog systems, and gradually becoming digitized. Network cameras and PC servers are now used for video recording in a fully digitized system.

7.2.6.2 Network Connectivity

Network connectivity is one of the most important key features of a safe city project and needs careful attention in assessment, planning, and implementation. This is the backbone of the system in which data travels from the surveillance systems to the data centers and control viewing centers. It is important to ensure that the provisioned connectivity is reliable and secure and not plagued with latency, jitter, packet loss, and performance. A combination of network technologies including leased lines, OFC network, terrestrial networks, wireless broadband, VSAT, and mobile networks are expected to be used to provide seamless connectivity for all surveillance. The provisioning of the network backbone should also ensure connectivity to the data centers and control rooms with scalable capacities to allow for expansion in the future.

7.2.6.3 Data Center

Data centers are the heart of any surveillance-based safe city projects. The data center acts as a warehouse for the data collected from the surveillance sensors. The data center is also responsible for providing continuous, real-time data to the command viewing centers for seamless, efficient, and effective operations. Generally, a primary and a secondary data center are established to ensure that the operations remain uninterrupted even if one is down. This center hosts all the applications that are required by the agencies to operate systems such as the video management software and the analytics application (VMS, VA), the Automatic Number Plate Recognition (ANPR) application, and the automatic vehicle classification. Appropriate space is provided for storage as well as retrieval of the digital information captured by the system. The design of a data center for a safe city primarily depends on the type of operations that are envisaged by the security agency in a safe city project and the type of processing required on these feeds: indexing, matching in DB, pattern analysis, GIS mapping, video analytics, facial recognition, etc.

7.2.6.4 Command Viewing Centers

A Command Viewing Center (CVC) is an infrastructure that accesses the collated and integrated information available at the data center such as incident video feeds. CVC allows the collation of information, thus helping in the analysis of data for quicker decision making. CVCs will be equipped with an intelligent operations capability to ensure integrated data visualization, real-time collaboration, and deep analytics that can help the agencies prepare for problems, coordinate and manage response efforts, and enhance the ongoing efficiency of city operations. The Graphical User Interphase (GUI) available at the CVC will equip users to take decisions by using the real-time and unified view of operations. Cities can rapidly share information across agency lines to accelerate problem response and improve project coordination. A CVC assists in leveraging information available with all the city agencies, thus allowing the management to make efficient and informed decisions. Furthermore, the center helps in anticipating the challenges and minimizing the impact of disruptions. A CVC will provide a city-wide GUI for visually depicting the video feeds and other sensor data. The GUI will also provide an overall status of the various city operations and its functions. The drill-down capability of such a dashboard will allow the operational users and decision makers to explore the underlying detailed status information to a depth relevant to their role.

The viewing center will have a GIS map of the city giving the status of the area of interest to the agency. Multiple map layers may depict equipment or other assets, events, weather, and positions of resources available to the city operations or boundaries of designated areas. Cross-agency collaboration supports messaging between operators at the control center, response units at the incident sites, and other agencies, with the aim of reducing the response time, sharing information effectively, and enabling collaborative decision making in a controlled and assured environment. It comprises a set of tools such as emergency call response systems and call dispatch systems to support immediate communication between all users and supporting agencies. Incident management capability is achieved through tools that assist in detection and management of incidents. The toolset enables the commanders as well as the executive staff to actively manage all the security aspects of the city since these tools provide real-time information of incident detection, incident correlation, and incident response. All of the capabilities of the CVC put together (the user interface, the GIS maps, the integration of application data, advanced analytics, and incident management) provide the shared situational awareness required to enable the city operations staff and supporting agencies to synchronize and prioritize. The operator will be trained and provided with the standard operating procedures for responses to such incidents and emergency situations for effective crisis management. The control center will be designed to enable all back-office operations which will be closely integrated with the command and control center operations.

7.2.6.5 Collaborative Monitoring

A key enabler for safe city is the aspect of collaborative monitoring. In Indian cities where every establishment, government or private, has realized the necessity to secure its infrastructure and establish surveillance, monitoring, and incident response systems, it is important that the data gathered by these agencies is shared among them. Government agencies such as the aviation and transport department are already deploying onboard surveillance systems by provisioning CCTV-based surveillance on public buses and bus stands, metros, railway stations, and airports. These systems under collaborative monitoring can conveniently share their data in real time with the security agencies of the city. Similarly, live feeds from CCTV systems deployed by private establishments such as malls, business parks, and entertainment houses can be provided to the CVC of the city where the security agency can make effective use of the information. Many cities across the world have surveillance systems deployed by multiple public and private establishments. These cities are using the collaborative framework to receive video feeds from these systems to ensure real-time responses.

7.2.6.6 Change Management and Capacity Building

The change management and capacity building programs form an integral part of the safe city project. These initiatives will acquaint the stakeholders to the proposed system and its associated processes. Furthermore, it will motivate, train, and empower the security agency officials to adopt revised methods of working and appreciate the resultant benefits. Change management will keep every stakeholder informed about the changes in the process flow and information management systems. It will empower the officials with the necessary skill and attitude in order to facilitate them in performing their duties in a more effective manner.

7.2.6.7 Safe City Drivers in India

In order to address the complex security challenges and to develop a data-driven decision-making approach, an illustrative evaluation framework of a safe city project is adopted that offers a detailed logic model for implementing the activities that will lead to the intended outcomes of reduced crime and faster response. State agencies can, however, evaluate their need of implementation of safe city projects on the basis of many other parameters. There can be many factors such as the prevailing crime rates, the extent of urbanization, threats from terrorism and disasters, the socioeconomic importance of an area, literacy rate, political importance of a particular place, etc., that can accentuate the need for adopting a safe city project. However, factors such as Internet penetration, network infrastructure, and the extent of industrialization aid in the successful implementation of such initiatives.

7.3 POLICY MAKING AND FRAMEWORK DEVELOPMENT

Public policy refers to the actions taken by the government and its decisions that are intended to solve problems and improve the quality of life for its citizens. At the federal level, public policies are enacted to regulate industry and business, to protect citizens at home and abroad, to aid state and city governments and people such as the poor through funding programs, and to encourage social goals (Stuart, 1990).

7.3.1 Process of Policy Making

A policy established and carried out by the government undergoes several stages from inception to conclusion. These are agenda building, formulation, adoption, implementation, evaluation, and termination (Fig. 7.2).

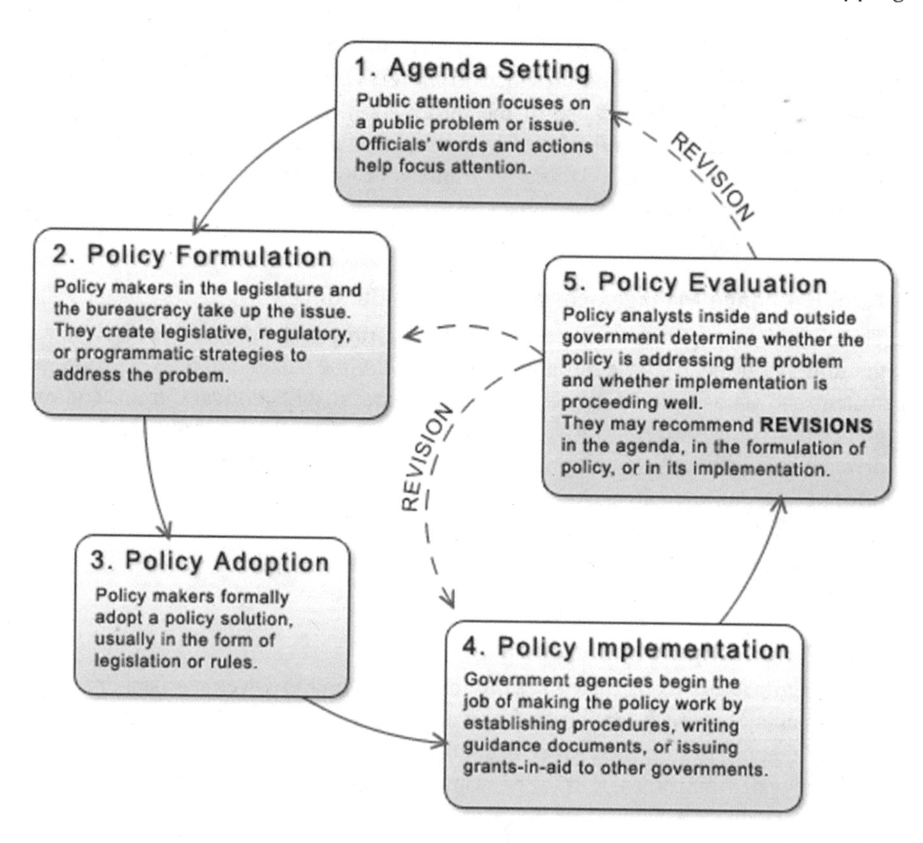

FIGURE 7.2 Stages of Policy Making.

7.3.1.1 Agenda Building

It becomes very essential to build an agenda before the actual policy making commences. For example, illegal immigration, has been going on for many years, but it was not until the 1990s that enough people considered it such a serious problem and felt that it requires increased government action. Another example is crime. Indian society has been tolerating a certain level of crime; however, when crime rose dramatically or is perceived to be rising dramatically, it became an issue for policy makers to address. Specific events can place a problem on the agenda. The flooding of a town near a river raises the question of whether homes should be allowed to be built on a floodplain.

7.3.1.2 Formulation and Adoption

Policy formulation refers to coming up with an approach to solving a problem wherein the legislature, the executive branch, the courts, and interest groups may be involved. Contradictory proposals are often made. The president may have one approach to immigration reform, and the opposition-party members may have another. Policy formulation then has a tangible outcome. A policy is adopted when parliament passes legislation, the regulations become final, or the Supreme Court renders a decision in a case.

7.3.1.3 Implementation

The implementation or carrying out of policy is most often accomplished by institutions other than those that formulated and adopted it. A statute usually provides just a broad outline of a policy. For example, the government may mandate improved water quality standards, but the Environmental Protection Agency (EPA) should provide the necessary details on those standards and the procedures for measuring compliance through regulations. Successful implementation depends on the complexity of the policy, coordination between those putting the policy into effect, and compliance.

7.3.1.4 Evaluation and Termination

Evaluation means determining how well a policy is being executed and adapted, which is not an easy task to monitor. People within and outside of the government typically use cost-benefit analysis to try to find the answer. In other words, if the government is spending x billions of rupees on a policy, are the benefits derived from it worth the expenditure? Cost-benefit analysis is based on hard-to-come-by data that are subject to different, and sometimes contradictory, interpretations. History has proven that once implemented, policies are difficult to terminate. When they are terminated, it is usually because the policy became obsolete, clearly did not work, or lost its support among the interest groups and elected officials that placed it on the agenda in the first place.

7.3.2 Framework Development

Theories are formulated to explain, predict, and understand phenomena and, in many cases, to challenge and extend existing knowledge within the limits of critical bounding assumptions. The theoretical framework is the structure that can hold or support a theory of a research study. The theoretical framework introduces and describes the theory that explains why the research problem under study exists. Here are some strategies to develop an effective theoretical framework:

1. Examine the title and research problem. The research problem anchors your entire study and forms the basis from which you construct your theoretical framework.
2. Brainstorm about what you consider to be the key variables in your research. Answer the question, "What factors contribute to the presumed effect?"
3. Review related literature to find how scholars have addressed your research problem. Identify the assumptions from which the author(s) addressed the problem.

4. List the constructs and variables that might be relevant to the study. Group these variables into independent and dependent categories.
5. Review key social science theories that are introduced in the course readings and choose the theory that can best explain the relationships between the key variables in the study.
6. Discuss the assumptions or propositions of this theory and point out their relevance to the research.

7.4 GIS FOR SOCIAL WELFARE

GIS is a key component in modernizing the Information Technology (IT) of many human and social service programs, paving their way to smarter and safer cities. By leveraging the data management, analysis, and visualization capabilities of GIS, social workers and other human service professionals (hereinafter referred to collectively as helping professionals) are empowered to understand community needs, measure environmental forces (including access to services), deliver services more efficiently, and detect fraud and abuse (Fig. 7.3). There is a strong historical and theoretical underpinning for the notion that place matters in human and social services.

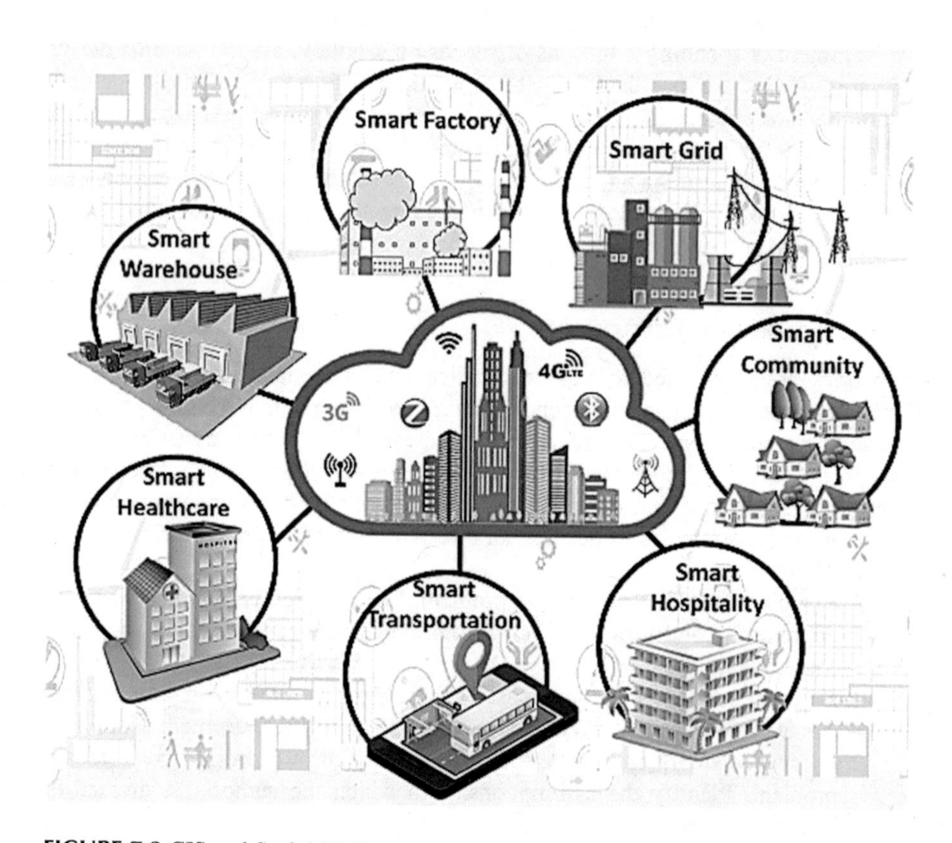

FIGURE 7.3 GIS and Social Welfare.

GIS helps illustrate unmet needs and overlapping or redundant services. Examples of GIS functions include thematic mapping (e.g., census data regarding children under the age of 5 living in poverty), geo-coding locations of existing service centers so they may be mapped and analyzed, and proximity analysis (e.g., distance buffering and constructing service areas) (ESRI, 2011). As Pamuk states, multilayer analysis through GIS is more powerful because it incorporates additional variables into decision making and can work with either vector or raster data.

Through GIS analysis, organizations address a major goal of programs: to have a proper match of services and resources to need. The ability of an agency or NGO to provide accessible services requires an understanding of the spatial distribution of the population it intends to serve as well as knowledge of the location and reach of existing services. Location is by no means the only factor in accessibility, but it is a major factor both in terms of physical accessibility of clients to a service site and distance for field staff to travel to clients.

Despite a history of leadership in mapping the impact of context on individual well-being, the use of GIS in social welfare remains rather uncommon and may lack sophistication. Social work has historically been criticized as reticent to adopt new technology and non-profits, where many social workers are employed, are thought to lag behind their business sector counterparts in the use of information technology. Further, social workers may be slow to adopt new technology because of limited training and education around the use of technology in schools of social work as compared to other disciplines. The same social workers who have called for the use of GIS and spatial approaches in social work have led the call for social work education and training to include exposure to GIS technology and opportunities to join interdisciplinary teams where these technologies might be employed. While the cost of software and training may be seen as barriers to more widespread adoption of GIS, some suggest that a more fundamental barrier maybe a new mind-set that marginalizes ecological and contextually focused social work practice.

In addition to the profession's slow adoption of technology, it has been suggested that the failure to embrace GIS may be a philosophical issue, reflective of social work's move toward more clinical approaches and away from contextual, community-oriented approaches. Despite our defining person in environment framework, we often fail to move beyond the social environment to understand the effects of the built and physical environments of the communities in which our clients reside. Qualitative GIS approaches may be a way for social workers to begin to get back to our roots and more deeply explore how place matters, with the assistance of technology already in use by many other disciplines.

7.4.1 GIS and Society

We have discussed in the earlier chapters how effectively GIS contributes in making our lives simpler and safer. The following points further discuss the usages of GIS applications in society as a whole.

7.4.1.1 Public Health

Public health seeks to address the health issues affecting people at the basic level or even at a larger national perspective. In the case of disease outbreaks, GIS is among the few systems that the government uses to do statistics. The availability of digital maps of any part of the world has over the years contributed to the huge ambiguity of GIS in the health sector. With these vigorous functionalities, better decision making is realizable to the fullest in health-related issues. GIS can also be used to determine the relationship between geographic features and disease origin contracted by persons living in those areas. Secure geographical information systems can be used to collect data from varying sources and perform statistical methods on the data.

7.4.1.2 Real Estate Development

Mapping in the survey, geographic statistics, and location addresses are other key features of GIS in real estate. A large collection of data is analyzed to help investors determine which areas would best prosper and work for real estate. It is done by considering population and demographic features available in a given location. Land with the best characteristics can be sorted and organized as needed, together with its characteristics. Robust geographic systems may be designed to even send updated information to clients who have indicated interest in real estate.

7.4.1.3 Urban and Rural Planning

GIS works wonders in urban planning. Be it creation of master plans, chalking down of transport networks, construction of railways, roads, or fly-overs, its applications are unimaginably helpful. Various ideas and innovations are being brought out by GIS to ensure the construction of world-class cities, towns, and rural areas with the least congestion. Various developments such as electricity, roads, railway, optic fibers, water and sewer lines determine how an urban or rural area is also in control of GISs.

7.4.1.4 Geo-demographics

Geo-demographics, a social science, involves studying a population of a given location. This population might belong to human beings or even animals. Geo-demographics specifically draws the idea about the number of people living at a given place over a certain time. Population maps can thus be analyzed using GIS.

7.4.1.5 Geographic History

Geo-demographics history involves the study of formation and activities of geographic features such as volcanic mountains, plateaus, hills, plains, and depressions. All of these features affected humans who existed then and must be studied to always be on the lookout today. History also involves the study and excavation of fossils. Fuels such as petroleum and human remains are part of the fossils studied. These remains are in most cases buried very deep below the earth, after long periods of geo-activities. It may thus be difficult to piece out useful information without the use of geographic systems such as GIS. GIS can be used to analyze their depths and characteristics before even an actual excavation by historians who want to do further research on them. GIS is thus essential in tracking the existence of humans in history.

7.4.1.6 Emergency Planning

GIS is very helpful in the areas of pandemic evaluation and planning, disaster management, and emergency planning. Natural disasters such as droughts and floods, fire tragedies, bridge collapses etc., require a quick response system and accurate mapping. Most governments have adapted GIS to help in analyses and mitigation of such calamities. GIS can be integrated with weather systems to offer predictions and useful statistics that help in decision making by the government. Satellite images of places encountering disasters can also be sent for analysis and forwarded to any base station on earth for processing by GIS. These systems would send help to enable navigation during evacuation operations of affected victims.

7.4.1.7 Criminology

Criminal investigation by intelligence services has received great help from GIS. GIS and remote sensing contributes in mapping of an area, generating LULC, highlighting pockets of crime, susceptibility zones, density profiling, etc., which facilitates the investigators to draw outcomes regarding crime arousal and trends in region. With the use of artificial intelligence, GIS effectively contributes to crime predictions in the future and crime patterns can be forecasted.

7.4.1.8 Transport Modeling and Support

Aviation is one of the fields in transport that receives maximum support from GIS. Transportation by air is a sensitive and delicate mechanism which needs to be very accurate in generation of route charts and to avoid plane crash tragedies. Location of aircraft can be tracked by the use of radars that are integrated with GIS, and this is very useful during reporting air travel emergencies that might require quick action. Structures such as those erected on roads, railways, bridges, tunnels, etc., are also planned and routed using GIS technologies in many countries. Slopes and drainage features are thus easier to study how they can in the future, or today, affect transport systems.

Thus it is appropriate to mention that GIS has promoted cost savings mechanisms and measures resulting in greater efficiency. By implementing GIS in logistics operations, there have been dramatic improvements in efficiency. With the use of GIS, customer service efficiency is increased by reducing the number of return visits to the same site and scheduling appointments more efficiently. GIS contributes to promoting marketing as well. Integrating the power of geography into marketing planning helps the organization respond to both customers and market needs. Through geo-marketing, GIS allows organization segments and profiles existing customers to improve acquisition and retention and find new opportunities, providing a competitive edge to succeed in today's volatile economy. GIS has also led education to the next level. It equips the education community with tools to develop a greater understanding of our world through geospatial data analysis. With GIS, students and teachers can integrate and evaluate data from diverse means to develop new theories and knowledge. This grooms the students to cope with the demands of the 21st-century workforce, whether involved in science, government, or business. Educative institutions such as libraries, museums, and universities are now using GIS for resource management, facilities management, and advanced research. Mapping technology of GIS has also aided the process. Search and rescue teams now have the advantage of a smooth communication with the base of operation (Tomaszewski, 2015). The biggest benefit of GIS has been improved communication. By improving communication, handling and containing the crisis situation has been given a huge boost. Rescue operations no longer have the problem of communication.

REFERENCES

Enhancing Urban Safety and Security: Global Report on Human Settlements. (2007) Un-Habitat, United Nations Human Settlements Programme, London • Sterling, VA.

ESRI. (2011) GIS for Human and Social Services Organizations: Place Matters in the Helping Professions. Esri, Redlands, CA, USA.

Markus, N. and Mitasova, H. (2002) Open Source GIS: A GRASS GIS Approach, The International Series in Engineering and Computer Science. Springer, New York, NY, Vol. 773, p. 406.

Parliamentary Debates: Official Report. (2006) Rajya Sabha Council of States Secretariat, National Informatics Centre, India, 208, pp. 20–21. https://rsdebate.nic.in/.

Saxena, A. K. (1997) Professionalism in Indian Police. APH Publishing Corporation, New Delhi, pp. 44–47.

Stuart, S. N. (1990) The Policy Theory and Evaluation: Concepts, Knowledge, Causes and Norms. Greenwood Press, New York, pp. 77–102.

Tomaszewski, B. (2015) Geographic Information Systems (GIS) for Disaster Management. CRC Press, Taylor and Francis, p. 42.

Walt, L. P. (2013) Predictive Policing: The Role of Crime Forecasting in Law Enforcement Operations. RAND Corporation, Santa Monica, CA.

Index

I

J

K

L

M

N

P

Q

R

S

T

U

V

Vehicle routing problems, 146

W

Z